WITHDRAWN FROM
TSC LIBRARY

A Field Guide and Identification Manual
for Florida and Eastern U.S. Tiger Beetles

INVERTEBRATES OF FLORIDA

Florida A&M University, Tallahassee
Florida Atlantic University, Boca Raton
Florida Gulf Coast University, Ft. Myers
Florida International University, Miami
Florida State University, Tallahassee
University of Central Florida, Orlando
University of Florida, Gainesville
University of North Florida, Jacksonville
University of South Florida, Tampa
University of West Florida, Pensacola

INVERTEBRATES OF FLORIDA

Guide to the Grasshoppers of Florida, by John L. Capinera, Clay W. Scherer, and Jason M. Squitier (2002)

A Field Guide and Identification Manual for Florida and Eastern U.S. Tiger Beetles, by Paul M. Choate, Jr. (2003)

A Field Guide and Identification Manual for Florida and Eastern U.S.

TIGER BEETLES

Paul M. Choate, Jr.

UNIVERSITY PRESS OF FLORIDA

Gainesville · Tallahassee · Tampa · Boca Raton · Pensacola · Orlando · Miami · Jacksonville · Ft. Myers

Copyright 2003 by Paul M. Choate, Jr.
Printed in China on acid-free paper
All rights reserved.
All photographs were taken by the author, P. M. Choate.

08 07 06 05 04 03   6 5 4 3 2 1

Choate, Paul Merrill, 1948-
A field guide and identification manual for Florida and Eastern
U.S. tiger beetles / Paul M. Choate, Jr.
p. cm.—(Invertebrates of Florida series)
Includes bibliographical references (p.).
ISBN 0-8130-2583-4
1. Tiger beetles—Florida—Identification.
2. Tiger beetles—East (U.S.)—Identification. I. Title. II. Series.
QL596.C56 C48 2003
595.76'2—dc21   2002035995

The University Press of Florida is the scholarly publishing agency
for the State University System of Florida, comprising Florida A&M
University, Florida Atlantic University, Florida Gulf Coast University,
Florida International University, Florida State University, University
of Central Florida, University of Florida, University of North Florida,
University of South Florida, and University of West Florida.

University Press of Florida
15 Northwest 15th Street
Gainesville, FL 32611-2079
http://www.upf.com

Contents

List of Illustrations   vii

Preface   xvii

Introduction   1
1. Why Study Tiger Beetles?   9
2. Habitats of Eastern Tiger Beetles   12
3. Florida's Geological History and Its Relationship to Animal and Plant Distributions   24
4. Species Criteria   37
5. Classification and Identification of Tiger Beetles   41
6. Species List: Eastern tiger beetles included in this book   45
7. Morphological Characters Used for Species Determination   47
8. Identification Key to Eastern Tiger Beetles   53
9. The Species of Eastern Tiger Beetles: Distributions and Habitats   61
10. Photographic Catalog of the Eastern Species of *Megacephala*   119
11. Photographic Catalog of the Eastern Species of *Cicindela*   121

Glossary   135

Works Cited and Suggested Readings   141

Index   195

Illustrations

Figure

1. Sample collection label to be attached to each specimen 5

Plates

1. *Cicindela gratiosa* adult 1
2. Head of tiger beetle (*Cicindela*) adult 2
3. Florida robber fly 4
4. Florida "bee fly," parasite of tiger beetles 4
5. Netting tiger beetles on Apalachicola River sandbar 5
6. Crossing the Blackwater River, Florida 7
7. Photographing *C. wapleri* on the Blackwater River, Florida 7
8. Clayton, Georgia: photographing *C. splendida* 8
9. Eglin AFB, Florida: roadside habitat 15
10. Route 24, Levy County, Florida: roadside cutting through oak scrub–sandhills 16
11. Josephine Creek, Highlands County, Florida: scrub habitat 16
12. Two miles west of Newberry, Florida: turkey oak scrub habitat 16
13. Apalachicola River, Florida 17
14. Blackwater River, Florida: sandbar habitat 17
15. Escambia River, Florida: shoreline habitat 17
16. Arkansas River, Pine Bluff, Arkansas: shoreline sandbar habitat 18
17. Connecticut River, Walpole, Vermont 18
18. Gulf coast, Alligator Point, Wakulla County, Florida 18
19. Larval burrows marked at Little Talbot State Park, Duval County, Florida 19
20. Little Talbot State Park, Duval County, Florida 19
21. Grassy Key, Monroe County, Florida: Gulf-side coral outcropping habitat 19
22. New Hampshire: habitat of *C. formosa generosa* 20

23. Maine: overgrown gravel pit, habitat of *C. longilabris*   20
24. New Hampshire: habitat of *C. tranquebarica* and *C. duodecimguttata*   20
25. Unicoi State Park, White County, Georgia   21
26. Clayton, Georgia: power line cut   21
27. Shell mound, Levy County, Florida: tidal salt marsh habitat   21
28. Two miles north of Horseshoe Beach, Dixie County, Florida   22
29. Blue Hill State Park, Suffolk County, Massachusetts   22
30. *C. marginata*   23
31. *C. dorsalis media*   23
32. *C. hirticollis*   23
33. Huntington Beach, Georgetown County, South Carolina: marsh area behind beach dunes   23
34. Head of *C. abdominalis* adult   47
35. *C. gratiosa* clypeus and labrum   48
36. *C. hirtilabris* clypeus and labrum   48
37. *C. marginata* male, mandibular tooth   48
38. *C. sexgutatta* head   48
39. *C. scutellaris unicolor* mouthparts   49
40. *C. formosa generosa* gena and lateral view of head   49
41. *C. formosa generosa*, dorsal view of head   49
42. Apex of female *C. hamata lacerata* elytron   49
43. *C. marginata* female, deflexed apex of elytron   50
44. *C. scabrosa* apical elytral microserrulations   50
45. *C. blanda:* identifying structures used in key   50
46. *C. blanda*   50
47. Components of elytral markings of tiger beetles   51
48. Typical complete markings of tiger beetles   51
49. *C. scutellaris unicolor*   51
50. *C. sexguttata*   51
51. Ventral view of tiger beetle   51
52. *C. pilatei*, decumbent setae on legs   52
53. Apical lunule of *Megacephala carolina floridana*   54
54. Apical lunule of *M. carolina carolina*   54
55. *M. carolina* (typical form)   62
56. *M. carolina* (typical form)   62
57. *M. carolina floridana*   63
58. *M. carolina floridana*   63
59. *M. virginica*   64

60. *M. virginica* adult scavenging at night   64
61. *C. duodecimguttata*   65
62. *C. duodecimguttata*   65
63. *C. formosa generosa*   66
64. New Hampshire: gravel pit   66
65. *C. formosa generosa*   66
66. Photographing *C. formosa generosa* larval burrow   67
67. *C. hirticollis* male   67
68. *C. hirticollis* female   67
69. *C. hirticollis rhodensis*   68
70. *C. hirticollis rhodensis*   68
71. *C. limbalis*   69
72. *C. limbalis*   69
73. *C. longilabris*   70
74. *C. longilabris* in "alert" position   70
75. Maine: habitat of *C. longilabris*   70
76. *C. nigrior*   71
77. *C. nigrior* adult   71
78. *C. patruela*   72
79. *C. purpurea*   72
80. *C. repanda*   73
81. *C. repanda* on Escambia River, Florida   73
82. *C. scutellaris unicolor*   74
83. *C. scutellaris unicolor*   74
84. *C. scutellaris unicolor*   74
85. *C. scutellaris unicolor*   75
86. *C. scutellaris lecontei*   75
87. *C. scutellaris unicolor*   75
88. *C. sexguttata* hiding in grass   76
89. *C. sexguttata*, showing granular elytra   76
90. *C. sexguttata*, immaculate form   77
91. *C. sexguttata*, maculate form   77
92. *C. splendida*, illustrating difference in color between elytra and head and pronotum   78
93. *C. splendida* with complete markings   78
94. *C. splendida* hiding in pine litter   79
95. *C. splendida* in "alert" position   79
96. *C. ancocisconensis*   80
97. *C. ancocisconensis*   80

98. *C. tranquebarica* 81
99. *C. tranquebarica* 81
100. *C. abdominalis* 82
101. *C. abdominalis* in "alert" position 82
102. Habitat of *C. abdominalis, hirtilabris,* and *scutellaris unicolor* 83
103. *C. highlandensis* 84
104. *C. highlandensis* 84
105. Josephine Creek, Highlands County, Florida: type locality of *C. highlandensis* 85
106. *C. highlandensis* 85
107. *C. marginipennis* 85
108. Adult *C. marginipennis* 86
109. Adult *C. marginipennis* 86
110. Connecticut River, Walpole, N.H.: habitat of *C. marginipennis* 86
111. *C. punctulata* 87
112. *C. punctulata* 87
113. *C. rufiventris hentzii* 88
114. *C. rufiventris hentzii* 88
115. *C. rufiventris hentzii* in "alert" position 89
116. *C. rufiventris*, dorsal view 89
117. *C. rufiventris* ventral view showing rufous abdomen 89
118. *C. rufiventris* on woodland path 90
119. *C. rufiventris* on eroded clay bank 90
120. *C. scabrosa* 91
121. *C. scabrosa* 91
122. *C. scabrosa* in "alert" position 92
123. Levy County, Florida: habitat of *C. scabrosa, hirtilabris,* and *scutellaris unicolor* 92
124. *C. trifasciata ascendens* 93
125. *C. trifasciata ascendens* 93
126. *C. dorsalis media* 94
127. *C. dorsalis media* 94
128. *C. dorsalis media*: mating pair 94
129. Larval burrows of *C. dorsalis media* marked at Little Talbot State Park, Duval County, Florida 95
130. *C. dorsalis saulcyi* typical markings 96
131. *C. dorsalis saulcyi* 96
132. *C. dorsalis saulcyi* 96

Illustrations · xi

133. Alligator Point, Wakulla County, Florida: habitat of *C. dorsalis saulcyi*   97
134. *C. severa*   98
135. *C. severa*, at night   98
136. *C. striga*   99
137. *C. striga*, at night   99
138. *C. striga*, at night   100
139. *C. togata*   100
140. *C. togata*, at night   101
141. *C. olivacea*   101
142. Grassy Key, Monroe County, Florida: habitat of *C. olivacea*   102
143. *C. viridicollis*, dorsal view   103
144. *C. viridicollis*, ventral view   103
145. *C. cursitans*   104
146. *C. unipunctata*   105
147. *C. unipunctata* in woodland path   105
148. *C. pilatei*   106
149. *C. pilatei*   106
150. *C. pilatei*: mating pair   106
151. *C. blanda*   107
152. *C. blanda*   107
153. Blackwater State Park, Santa Rosa County, Florida: habitat of *C. blanda*   107
154. *C. cuprascens*   108
155. *C. gratiosa*   109
156. Head of *C. gratiosa*, showing glabrous labrum   109
157. *C. gratiosa*   109
158. *C. hamata lacerata* male   110
159. *C. hamata lacerata* female   110
160. Tidal mudflat exposed at low tide: habitat of *C. hamata lacerata* adults   111
161. Larval burrow of *C. hamata lacerata* near Cedar Key, Florida   111
162. *C. hirtilabris*   112
163. *C. hirtilabris* clypeus and labrum   112
164. Mating pair of *C. hirtilabris*   112
165. *C. hirtilabris*   113
166. *C. lepida*   114
167. *C. macra*   114

168. *C. marginata* male   115
169. *C. marginata* female   115
170. *C. marginata*   115
171. *C. puritana*   116
172. Calvert Cliffs, Maryland: habitat of *C. puritana*   117
173. *C. puritana*   117
174. *C. wapleri*   118
175. *C. wapleri*   118
176. *Megacephala carolina carolina*   119
177. *M. carolina floridana*   119
178. *M. virginica*   119
179. *C. duodecimguttata*   121
180. *C. formosa*   121
181. *C. hirticollis* female   122
182. *C. hirticollis* male   122
183. *C. purpurea*   122
184. *C. splendida*   122
185. *C. limbalis*   123
186. *C. patruela*   123
187. *C. longilabris*   123
188. *C. sexguttata*   123
189. *C. sexguttata*   124
190. *C. sexguttata*, maculate specimen   124
191. *C. repanda*   124
192. *C. nigrior*   124
193. *C. scutellaris unicolor*   125
194. *C. scutellaris unicolor*   125
195. *C. ancocisconensis*   125
196. *C. tranquebarica*   125
197. *C. abdominalis*   126
198. *C. highlandensis*   126
199. *C. scabrosa*   126
200. *C. marginipennis*   126
201. *C. punctulata*   127
202. *C. trifasciata ascendens*   127
203. *C. rufiventris hentzii*   127
204. *C. rufiventris*   127
205. *C. dorsalis saulcyi*   128
206. *C. dorsalis media*   128

207. *C. dorsalis saulcyi*   128
208. *C. togata*   128
209. *C. severa*   129
210. *C. striga*   129
211. *C. olivacea*   129
212. *C. viridicollis*   129
213. *C. cursitans*   130
214. *C. unipunctata*   130
215. *C. pilatei*   130
216. *C. blanda*   130
217. *C. cuprascens*   131
218. *C. macra*   131
219. *C. wapleri*   131
220. *C. puritana*   131
221. *C. lepida*   132
222. *C. gratiosa*   132
223. *C. hirtilabris*   132
224. *C. marginata* male   132
225. *C. marginata* female   133
226. *C. hamata lacerata* male   133
227. *C. hamata lacerata* female   133

Maps

1. Florida counties   61
2. Florida distribution of *Megacephla carolina carolina*   62
3. Eastern distribution of *M. carolina carolina*   62
4. Florida distribution of *M. carolina floridana*   63
5. Florida distribution of *M. virginica*   64
6. Eastern distribution of *M. virginica*   64
7. Eastern distribution of *C. duodecimguttata*   65
8. Distribution of *C. formosa generosa*   66
9. Eastern distribution of *C. hirticollis*   67
10. Florida distribution of *C. hirticollis*   68
11. Eastern distribution of *C. limbalis*   69
12. Eastern distribution of *C. longilabris*   70
13. Florida distribution of *C. nigrior*   71
14. Distribution of *C. nigrior*   71
15. Eastern distribution of *C. patruela*   72

16. Eastern distribution of *C. purpurea* 72
17. Eastern distribution of *C. repanda* 73
18. Florida distribution of *C. repanda* 73
19. Florida distribution of *C. scutellaris unicolor* 74
20. Distribution of *C. scutellaris unicolor* 74
21. Eastern distribution of *C. sexguttata* 77
22. Florida distribution of *C. sexguttata* 77
23. Eastern distribution of *C. splendida* 79
24. Distribution of *C. ancocisconensis* 80
25. Eastern distribution of *C. tranquebarica* 81
26. Florida distribution of *C. tranquebarica* 81
27. Southeastern distribution of *C. abdominalis* 83
28. Florida distribution of *C. abdominalis* 83
29. Distribution of *C. highlandensis* 84
30. Known distribution of *C. marginipennis* 86
31. Eastern distribution of *C. punctulata* 87
32. Florida distribution of *C. punctulata* 87
33. Distribution of *C. rufiventris hentzii* 88
34. Eastern distribution of *C. rufiventris* 90
35. Florida distribution of *C. rufiventris* 90
36. Distribution of *C. scabrosa* 92
37. Eastern distribution of *C. trifasciata ascendens* 93
38. Florida distribution of *C. trifasciata ascendens* 93
39. Eastern distribution of *C. dorsalis media* 95
40. Florida distribution of *C. dorsalis media* 95
41. Distribution of *C. dorsalis saulcyi* 97
42. Florida distribution of *C. dorsalis saulcyi* 97
43. Eastern distribution of *C. severa* 98
44. Florida distribution of *C. severa* 98
45. Eastern distribution of *C. striga* 100
46. Florida distribution of *C. striga* 100
47. Eastern distribution of *C. togata* 101
48. Florida distribution of *C. togata* 101
49. Distribution of *C. olivacea* 102
50. Florida distribution of *C. olivacea* 102
51. Distribution of *C. viridicollis* 104
52. Eastern distribution of *C. cursitans* 104
53. Distribution of *C. unipunctata* 105
54. Eastern distribution of *C. pilatei* 106

55. Eastern distribution of *C. blanda*   108
56. Florida distribution of *C. blanda*   108
57. Eastern distribution of *C. cuprascens*   108
58. Eastern distribution of *C. gratiosa*   110
59. Florida distribution of *C. gratiosa*   110
60. Distribution of *C. hamata lacerata*   111
61. Florida distribution of *C. hamata lacerata*   111
62. Distribution of *C. hirtilabris*   113
63. Florida distribution of *C. hirtilabris*   113
64. Distribution of *C. lepida*   114
65. Eastern distribution of *C. macra*   114
66. Distribution of *C. marginata*   116
67. Florida distribution of *C. marginata*   116
68. Distribution of *C. puritana*   117
69. Distribution of *C. wapleri*   118
70. Florida distribution of *C. wapleri*   118

Table

1. Geological timescale   36

Preface

Tiger beetles are easily recognized, colorful, and challenging to collect. Once a tiger beetle is collected, even the general collector may want to identify the species. That is the purpose of this publication, to enable the nonspecialist to identify species of tiger beetles found in the eastern United States.

This book is the culmination of several decades of sporadic fieldwork conducted in the eastern United States. I encountered my first tiger beetle while making a general insect collection at the University of Vermont in 1967. My roommate, Robert Davidson, had convinced me to take a field zoology course offered by Ross T. Bell. One of the larger components of the grade consisted of a collection. It was spring and, in Vermont, insects were not terribly abundant. We were stripping bark off dead tree limbs when I chanced upon a brilliant green, white-spotted beetle. It was an overwintering adult of the tiger beetle *Cicindela sexguttata*. I was struck by the brilliant color, the ferocious appearance of the large mandibles, the large eyes, and the quick reaction to being uncovered—flight into the adjacent pasture. I found additional specimens and collected several. After identifying these with the help of Ross Bell, I wanted to learn more about these interesting beetles. I found out that Vermont had a number of species, so I set forth to find and collect them; however, I was successful at finding only a few additional species, and unable to find the more desirable or rare ones. Soon afterward I departed for a four-year tour in the Air Force. My first station of any duration was at Austin, Texas. I spent two years searching for some of the many species of tiger beetles resident there. Thanks to my good friend Donald A. Wilson I was put in touch with Ron Huber, the editor of the journal *Cicindela*, devoted exclusively to tiger beetles. After two years in Texas I was assigned to Vietnam for a year, where I occasionally collected tiger beetles around lights at barracks and other facilities. I managed to have some of these identified by specialists at the Bishop Museum in Hawaii. I returned to finish my military tour at Sumter, South Carolina, where I tried to collect as many species as were recorded from that state. The literature was sparse, and identification a little more difficult. I then returned to finish my undergraduate work at the University of New Hampshire, majoring in entomology.

During two years at New Hampshire I was paid to collect beetles, to build the department's collection. Marcel Reeves encouraged me to work on the ground beetles as well as the tiger beetles. I managed to meet with and get to know P. J. Darlington. People working on the New England tiger beetle and ground beetle fauna included my friends Don Wilson, Robert Davidson, Paul Miliotis, Gary Dunn, Marcel Reeves, Wally Morse, and Ross Bell. After finishing my undergraduate degree, I was accepted into a graduate program in entomology at the University of Florida.

When I first arrived in Florida in 1975 I estimated that with a few summers of collecting the tiger beetle fauna would be easily figured out and ready for publication. Best-made plans often go astray, and finding all recognized species of tiger beetles proved a real challenge. Part of this challenge, resulting from recognizing habitat preferences, lay in finding these habitats before they could be developed, to disappear forever. Another major obstacle was the lack of collection data, species identification literature, and local experts who knew where to find the many species known to occur in Florida.

Remarkably, even a state such as Florida, heavily collected by northern entomologists for decades, yielded surprises. The discovery of a new species in Highlands County led to cooperative work with the Nature Conservancy to document the range and habitat of this species, ultimately resulting in its placement on the list of endangered species of invertebrates. There have been many other surprises. Perhaps second to the discovery of a new species was the discovery of a site in Dixie County, south of Steinhatchee, that was found to contain many species thought to occur only in more southern Florida, even into the Keys.

Florida suffers from rapid development and ever increasing population. The first habitats to be developed are the higher, drier sandy areas, and these are also the preferred habitats of many of Florida's tiger beetles. Therefore, many populations have become extinct. Coastal species such as *Cicindela dorsalis media* and *C. dorsalis saulcyi* disappear wherever vehicles are allowed to drive on the beach. *C. highlandensis*, described from central Florida (Choate 1984), may be found as small, isolated populations in Polk and Highlands counties. This species may have actually disappeared from its type locality (place from which it was first collected and described) because of over-collecting and habitat destruction.

While many people do not believe that a species can be eliminated by over-collecting, there are suggestions that this may well be the case with tiger beetles. Some species are poor fliers occupying a small, restricted area and easily collected. For these, it may be possible to reduce a population sufficiently to result in its elimination from that site. Others, such as the coastal species *C. dorsalis*, have been eliminated from miles of Florida coastline, not by over-collecting but by destruction of larval habitat by vehicular and pedestrian traffic. Larvae occur at the high tide line in many areas, the very place cars and people traverse because of the compactness of the sand.

The fragile nature of habitats required by many species remains to be documented. This alone makes these colorful insects worthy of study. Their keen eyesight, quick reactions, and rapid flight make them challenging to approach, and have resulted in a large following of collectors, both amateur and professional. The journal *Cicindela* provides reporting of new discoveries for this specific group. It is because of this widespread popularity that I decided to complete this book. Many non-entomologists would be quick to study these beetles if they had a means to identify specimens and access to pertinent literature to permit more extensive studies. These are the goals of this book.

I cannot say with certainty that we have discovered and collected all species of tiger beetles that occur in Florida. Undoubtedly there remain new discoveries to be made. In anticipation of this, I have included species from other eastern states that may be found in Florida, as well as other eastern species that most likely won't be found in Florida but that may be collected by northern tiger beetle fans. This was done for the sake of completeness. This book then includes all described species east of the Mississippi River and several varieties (subspecies), bringing the number of forms treated here to 45. There are presently approximately 112 recognized species of tiger beetles in the entire United States. While there have been several regional treatments of tiger beetle faunas, no single one treats all of the eastern species.

Florida is unique in many ways. One of these is its close proximity to Cuba. Until recently one species of tiger beetle known from Cuba was also known to occur in the Florida Keys, *Cicindela olivacea*. This number has increased to two with the collection by a USDA entomologist of *C. viridicollis*, also from the Keys. While this record is based on a single specimen, the small size of the species, as well as its secretive behavior, may mean that it has been overlooked all this time. The specimen may also represent a stray individual. Such records are the reason that we can never be absolutely sure we have all the known species from any one state or region.

This manual is intended to aid in the identification of tiger beetles. More than that, however, I hope the user will find reason to study these marvelous insects and their behavior, and to use them for conservation and educational studies.

I would like to acknowledge a few individuals who made this book possible. No words are adequate to express my thanks for the continued love and support of my wife Angela, who has continued to hope and finally believe that this would be finished. Friends and fellow collectors Donald Wilson, Robert Davidson, Scott Gross (deceased), Lloyd Davis, Jr., Mike Thomas, and Gary Dunn shared many fun times with me collecting and conversing about tiger beetles. Don Wilson spent many hours teaching me the habitats and collecting techniques for northern tiger beetles, including a trip to northern Maine in search of *Cicindela longilabris*. Of course, no acknowledgment is complete without thanks to those people responsible for bringing me into the world, Mom and Dad. While they may not always have understood my fascination with beetles they never tried to dissuade me from my pursuits. Thanks also to the anonymous reviewers of this manuscript, always a thankless job, and to my friends Bill, Larry, and Rocke for sharing many humorous moments along the way, and to all of our friends who supported us during recent difficult times. They were always there when we needed them. Thanks to Bruce and Joyce, for special moments, special trips, and special experiences. Thanks to friends and family who wonder at the merits of a book about beetles, but remain curious enough to see what this is all about. Finally, many thanks to all other fellow Cicindelophiles who have generously offered support and collecting information over the years.

Introduction

Carabidae—Subfamily Cicindelinae
   Common name: Tiger beetles
   Cicindelidae Latreille, 1804; Collyridae Hope, 1838;
      Ctenostomidae Castelnau, 1835; Mantichoridae Castelnau,
      1835; Megacephalidae Castelnau, 1835.

Tiger beetles—often brilliantly colored, very active, predaceous—are popular insects. Historically they have been classified as the family Cicindelidae, separate from the ground beetle family Carabidae, mainly because of their recognizable features and appeal to collectors, but there is every indication that they should be considered a subfamily of Carabidae. Many workers have accepted the close relationship of this group to the subfamily Carabinae; this association is in agreement with larval and adult characters and has been maintained throughout the history of the classification of tiger beetles. At this point it is a matter of personal opinion whether this group should or should not be recognized as a separate family. There are no set rules for the determination of family rank, but if the families of beetles are to be delineated consistently, tiger beetles should be placed within the ground beetles.

What is a tiger beetle?

Tiger beetles are separated from other ground beetles (Carabidae) by a combination of the following characters: clypeus laterally extending beyond antennal insertion (plate 2), both spurs of protibiae being apical, and larvae having dorsal hooks on fifth abdominal tergite.

Plate 1. *Cicindela gratiosa* adult.

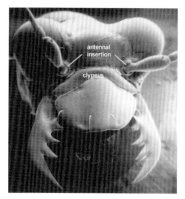

Plate 2. Head of tiger beetle (*Cicindela*) adult.

Description: (Plate 1) Conspicuous eyes, long legs, frequently with brilliant colors, characteristic shape, rapid locomotion, and nervous activity are characteristics of this group. Shape various, body elongate or round, with large head and prominent eyes; length ranges from less than 10 mm (*Cicindela*) to 50 mm (*Amblycheila:* United States) or 70 mm (*Manticora:* South Africa); color often with iridescent markings; some genera uniformly black; vestiture usually sparse; long, stout setae, especially on legs and mouthparts.

Head variously shaped, prominent, not elongate; antennae filiform, long, slender, with eleven antennomeres; mandibles stout, toothed, sometimes very large and prominent, much larger than the head; other mouthparts large; eyes varying from very small to very large and prominent.

Pronotum variously shaped, narrower than elytra, usually cylindrical, without lateral margins, or rarely with lateral margins; legs long and slender, fitted for running; tarsi 5-5-5; typical carabid-type wing; elytra covering the abdomen, sometimes connate, rounded behind, sometimes with lateral margins; hind wings sometimes absent.

Abdomen sometimes narrow, with elytra covering the sides, often as broad as elytra, with four to six visible ventral sternites. Male genitalia highly modified, carabid trilobed type parameres with a separate pars basalis (unlike that found in other carabids), entirely separate from the base of the penis and forming a lateral connection between the parameres at a midway point along them. Larvae distinguished by two or three pairs of hooks present on the tergum of the fifth abdominal segment.

Ecology. Pearson (1988) and Pearson and Vogler (2001) summarized the biology, systematics, physiology, behavior, biogeography, genetics, and anatomy of this group. Tiger beetles are predaceous and often run very rapidly over the ground, frequenting sandy banks of streams, edges of roadways, paths, and other such sunny places. Many species are rapid flyers. In spite of their often brilliant metallic colors they are difficult to see because of their quick movements. Unlike most other beetles, these often require a net for collection. Some desert-inhabiting species can be collected in abundance by hand. Pitfall trapping has also proven a successful technique (Boyd 1985; Dunn 1980b; Franklin 1988) for some species. Many species are more easily collected at night with such attractors as blacklights or mercury vapor lights, but overzealous collecting at lights of many of the more elusive species has led to the demise of some populations. Several species are on the verge of extinction, with over-collecting appearing to be as responsible as habitat destruction.

The exotic genera *Mantica* and *Manticora* inhabit deserts and steppes. A few exotic species of *Cicindela* live in termite nests. The tropical tribes Ctenostomatini and Collyrini are arboreal, searching out their prey among the branches of trees and bushes.

Species of *Cicindela* are both diurnal and nocturnal. Many species prefer dry sunny areas, but others are known to prefer semiaquatic sites. Some species, while active both during daytime and at night, appear to accomplish most of their mating at night (for example, *dorsalis*, *hamata*, *marginata*). Many are strong flyers, but some have fused elytra and are incapable of flight (for example, *pilatei*, *belfragei*). Others appear reluctant to fly even though flight wings are present (for example, *unipunctata*, *cursitans*).

Larvae for most known species live in vertical tunnels, holding themselves near the top of the burrow with the aid of dorsal hooks on the fifth abdominal segment. If suitable prey comes within range of a tunnel entrance, the larva seizes it and drops down into the burrow to feed. The abdominal hooks also serve to anchor the larvae in its tunnel if a prey is too large to subdue. Larvae of the Collyrini and Ctenostomatini (*Neocollyris* and *Ctenostoma*) are unusual in that they develop in rotting logs and branches or standing vegetation (Zikán, 1929) rather than in burrows in the ground. Wilson (1974) demonstrated survival from flooding by larvae of three species of northeastern tiger beetles. Hamilton (1925) described larvae of many Holarctic species; since then, other tiger beetle larvae have been described (Beatty and Knisley 1982; Gilyarov and Sharova 1954; Horn 1878; Kaulbars and Freitag, 1993a; Knisley and Pearson 1984; Leffler 1980b, 1985b; Putchkov 1990, 1993, 1994a,b, 1995; Putchkov and Arndt 1994; Putchkov and Cassola 1991; Shelford 1908; Willis 1967, 1980).

Predators and parasites

Tiger beetles are subject to predation by a number of vertebrates, especially birds, lizards, toads, and frogs. Larochelle (1974b) listed amphib-

Plate 3. Florida robber fly (Diptera: Asilidae).

Plate 4. Florida "bee fly," parasite of tiger beetles (Diptera: Bombyliidae).

ian and reptile predators of tiger beetles. Among invertebrate predators are robber flies (Diptera: Asilidae), dragonflies, spiders, and some species of ants. Parasites include bee flies (Diptera: Bombyliidae) and wasps of the family Tiphiidae. Graves (1962) reported predation of *Cicindela repanda* by the dragonfly *Aeschna interrupta*. Wallis (1961) reported large robber flies capturing adult tiger beetles on the wing (plate 3).

Shelford (1908) reported 7 percent of the larvae of *C. scutellaris lecontei* parasitized by the dipteran *Spongostylum anale* Say (Diptera: Bombyliidae) (plate 4). Frick (1957) observed eggs being deposited in larval burrows by a "small black bombyliid fly."

Collecting and preserving adult tiger beetles

Tiger beetles can be fun and quite challenging to collect because of their good vision, rapid running, and quick flight response.

Some species are more easily collected with pitfall traps, made from cans placed into the ground. This works if you will be in an area for some time and can check the traps on a daily basis. Most often collecting is done using an aerial net, but specimens are quick to run out from under the net, or to bury themselves in the sand and disappear before your very eyes.

Many species are attracted to artificial lights at night. Running an ultraviolet light may attract large numbers of individuals of several species, including, in Florida, *dorsalis media, dorsalis saulcyi, togata, marginata, hamata lacerata, severa, striga, gratiosa, hirtilabris, punctulata, blanda, wapleri,* and *olivacea*. There are many species that are not active at night and thus are seldom if ever collected at lights. In Florida these include *repanda, tranquebarica, sexguttata, hirticollis, abdominalis, scabrosa, highlandensis, scutellaris unicolor, formosa,* and *nigrior*.

Plate 5. Torreya State Park, Liberty County, Florida. Scott Gross netting tiger beetles on Apalachicola River sandbar.

Tiger beetles, though brightly colored when alive, tend to exude grease that dulls their colors after being killed. Acetone or ethyl acetate make effective degreasing compounds. Soak specimens for several days until colors are restored. If large numbers of specimens are cleaned, several changes of cleaning solution may be necessary. Both of these compounds are extremely dangerous as they are flammable and toxic to breathe. Use in a well ventilated area and avoid sparks or fire.

Specimens should be pinned through the right elytron, near the shoulder area. A label with collection data (fig. 1) should be pinned below each specimen. Collection data should include location, date of collection, collector, habitat, and collection technique, and determination if identified. More than one label may be used. The identification and person who made the determination should be attached on a second label, with date of determination included. Labels should be aligned parallel with specimen's body and not extend beyond body extremities.

```
USA: Florida; Alachua Co.
Gainesville, 27-VII-2001
pitfall trap, Turkey Oak
P. M. Choate collr.
```

Figure 1. Sample collection label to be attached to each specimen.

Collecting and preserving tiger beetle larvae

Most larvae may be found in the same habitat as the adults. Some larvae have vertical or horizontal burrows that extend several feet. For these, the only way to collect them is to use a long grass stem. Push the stem into the burrow as far as possible, then dig parallel to the stem until the larva is encountered. A more entertaining way to collect larvae from shallow burrows is by "doodling." This is an especially effective way to collect *Megacephala* larvae, which are seldom more than ten to twelve inches deep. Using a straight blade of grass, push the blade to the bottom of the burrow and tap it sideways a few times. If the burrow is occupied you will see the blade twitch as the larva grabs it. Give a quick yank and frequently the blade of grass will come out with the larva still grasping it. Larvae may be reared in glass tubes, or killed in hot water and preserved in 70 percent alcohol.

Locating and Photographing Tiger Beetles

Habitat recognition and knowledge of the seasonality of each species are critical to locating tiger beetles (see chapter 3, Habitats of Eastern Tiger Beetles). Once one has located the habitat for a particular species, the following elements determine whether or not specimens will be seen:

Temperature—once surface temperatures are too hot, individuals take cover, become inactive.

Time of day—midday is least optimal in the south; mid-morning and mid-afternoon for diurnal species; one hour after sunset for nocturnal species.

Cloudiness—diurnal species take cover when clouds obscure the sun. Individuals will "freeze" when a cloud passes in front of the sun, then resume activity as soon as the clouds pass.

Because of the active, alert behavior of adult tiger beetles, successful photography in the field requires patience, knowledge of the behavior of the species being hunted, a partner to distract or herd potential subjects, and luck. Perseverance usually pays off, but many hours may be spent chasing individuals until one remains quiet long enough for you to get within eight inches to photograph. Keeping a low profile seems to work well, and getting used to crawling on hands and knees over rocks and other debris is part of the challenge. A properly exposed, in-focus slide of a live tiger beetle in its habitat is something to be proud of! Equipment used by me is listed below.

Plate 6. Blackwater River, Florida. Crossing to get to habitat of C. *wapleri* and *blanda*.

Plate 7. Blackwater River, Florida. Photographing C. *wapleri* and *blanda*.

Field photography—Equipment:
105mm MicroNikkor lens, 1:1 magnification
Synchronized at 1/125 sec.
Nikon FM2 camera body
Lepp Bracket with 2 strobe units
ASA 100 slide film (Fujichrome)
f/11–f/16, depending on subject coloration
working distance: at 1:1, approx. 9 inches (the challenge!)
Studio photography—Equipment
Microscope camera
ASA 160 Tungsten Kodak Ektachrome film
Fiber optic lighting

Plate 8. Clayton, Georgia. Photographing *C. splendida*.

1

Why Study Tiger Beetles?

Tiger beetles may be key indicators of unique habitats containing previously undocumented or unrecognized species of plants or animals. Tiger beetles are sufficiently large to be observed from a distance with binoculars. Individuals may be monitored and their activity recorded, and with the use of field guides such as this it is possible to recognize species being observed. Thus it becomes possible to record unusual sightings, new distribution records that may contribute to the knowledge of a particular species. The casual observer may ask why that species occurs where it does, and may also begin to wonder what other organisms are tied into that same habitat and also going unnoticed. When Vermont naturalists discovered populations of *C. marginipennis* they also found that the same sections of river also harbored a rare species of mussel as well as endangered species of plants.

The same may be said for other rare or endangered species of plants or animals. Such organisms are unlikely to be the sole unusual entity in their habitat. Insects do not occur in their preferred habitat by chance. Understanding their requirements will lead to an understanding of complex, multidimensional relationships shared by all other plants and animals in a particular habitat. Tiger beetles are excellent teaching aids for science classes, conservation studies, and nature enthusiasts. As such, tiger beetles make good pointers to other equally diverse animals. Therefore, the study of tiger beetles opens the door to more knowledge about our environment and the organisms that share it.

When asked what was so special about the Puritan tiger beetle, biologist for the Maryland Heritage Program Rodney Bartgis replied, "They're an indicator of how well the shoreline ecosystem is working. If something's going wrong, tiger beetles are the first to change" (Surkiewicz 1993).

Tiger Beetles and Conservation Issues

The most important factor affecting tiger beetle distribution is habitat. Destruction of habitat, through development, damming of rivers, vehicular traffic on beaches, and heavy pedestrian traffic, and even over-collecting have been shown to reduce or eliminate populations of tiger beetles. Three examples are discussed here.

The puritan tiger beetle (*Cicindela puritana*) used to occur along stretches of the Connecticut River from Claremont, N.H., to near the mouth of the river in Connecticut (Leng 1902c). Today, most suitable habitats have been covered by dams and only a few suitable patches of banks and shoreline persist. These are now protected, and efforts are underway to permanently preserve the habitat. New England populations prefer wide sand deposits along big rivers or narrow beaches along rivers with clay banks (U.S. Fish and Wildlife Service 1990).

The range of the cobblestone tiger beetle (*Cicindela marginipennis*) has been studied and documented in New England (Dunn and Wilson 1979; Dunn 1981). There are only twelve known spots where this insect occurs, on cobblestone sandbars along the Connecticut River. *Cicindela marginipennis* attracted the attention of the media (Donnelly 1986) when Vermont legislators announced plans to protect it, making it the first-ever protected insect, along with its habitat along the banks of the Connecticut River. The state of Vermont has made preservation of this species a priority, protecting sections of habitat that are found on state land. In addition, six other species of beetles have been recognized as endangered by top state environmental officials.

Educators have found that by publicizing this species children have become more aware of the importance of habitat preservation. In New Hampshire, Plainfield townspeople named it the town insect; T-shirts were printed with a picture and the scientific name of the tiger beetle. At a Fourth of July parade a Volkswagen was disguised as a tiger beetle. Sara Townsend of Plainfield composed a tiger beetle song to the tune of the *Battle Hymn of the Republic*. "The kids think it's great, and they learned things they never knew before. They never knew about tiger beetles, but then again, nobody knew about them," said one of the Plainfield residents. As a result of the publicity and studies by environmentalists, a dam planned for construction below Plainfield was halted. State fish and wildlife officials posted access spots to the river to prevent all-terrain vehicles from driving on the cobblestone beach occupied by this insect. Charles Johnson, State of Vermont naturalist, stated in the newspaper article, "We tend to think [of] something big and glamorous as worthy of protection, and [of] small things as not worthy. . . . But there are pretty important things that are small. Tiny things can have great importance. . . . A lot of people like tiger beetles" (Donnelly 1986).

Vermont consultant for the Nature Conservancy Marc Desmeules stated, "People have jumped on this bug because it's something they can relate to. Endangered species are usually tigers or leopards you read about or see on television. This is right in our backyard. . . . The state is sticking their necks out on this one, to protect insects," Desmeules said. "If you talk to your average sportsman, or your average hunter, and who knows, they [state officials] may be opening themselves up to a lot of ridicule" (Donnelly 1986).

"We'll probably get teased about this," said Lawrence Garland, Vermont's assistant director of wildlife. "But we have a commitment to anything that is endangered. I think this shows the department isn't all a hook and bullet crowd" (Donnelly 1986).

The Florida highlands tiger beetle (*C. highlandensis*), discovered in central Florida and described in 1984 (Choate 1984), typifies the problems of rare or endangered species. Researchers from the Nature Conservancy have located and documented many small populations in Polk and Highlands counties, in central Florida. The persistence of this species is dependent on the preservation of the sand scrub habitat it occupies (Plate 105). This beetle is a relatively weak flier, and over-collecting at the type locality has so reduced adult populations that it may no longer exist there. This species is now found only along the Lake Wales Ridge of central Florida, a region noted for its endemism in many groups of animals and plants. This area is also under extreme pressure by developers, and by the citrus industry for land for citrus groves.

2

Habitats of Eastern Tiger Beetles

Tiger beetles are most often found on sandy soil. They frequent river sandbars, gravel pits and sandpits, coastal beaches, and roadside cuts where exposed soil exists. However, there are also many species that are found in other habitats, including salt marshes, rocky outcropping (*hentzii*), high mountain grassy areas (*splendida, purpurea*), sandy trails through woods (*patruela*), abandoned stone quarries, overgrown paths, clay banks, algae-encrusted mudflats, and coral outcropping. There are presently no known arboreal tiger beetle species in the United States.

Habitat refers to the sum of environmental conditions that are favorable for a tiger beetle to exist. In the northern states tiger beetles are most easily located in sandy areas. These are uniquely distributed in patches and fairly difficult to locate without the help of local collectors. However, when looking for tiger beetles in the southern states, especially in a state such as Florida, one finds almost every location to be sandy. Then other criteria come into play in habitat preferences of a particular species.

Some species of Florida tiger beetles are restricted to peninsular Florida even though the habitat they occupy in the peninsula is also present in the Panhandle. Other species occur only in the Panhandle and do not extend their range into peninsular Florida. Such distribution patterns are generally accepted as reflecting past geological history of Florida, when the peninsula was separated from the mainland by the Suwannee Straits (see chapter 3, Florida's geological history). Within peninsular Florida, distribution patterns on disjunct sand ridges and "islands" reflect geological events that saw peninsular Florida split into several smaller islands.

In Florida one cannot easily locate distinctly different habitats and find particular species of tiger beetles without some study of the native soil types and vegetation distributions. Several authors have applied restricted names to habitat types, defined by soil types and associated plant communities. I prefer to use more descriptive and generalized habitat names. For the purposes of this manual such tiger beetle habitats include roadsides, sand scrub, rivers and stream banks and sandbars (riparian), seashores and coastal dunes (littoral), gravel pits, coral outcroppings, woodland paths and roads, eroded clay banks, salt marshes, and alkali mudflats. Examples of these habitat types are illustrated here, along with listings of the species commonly found there.

Many of the eastern species not occurring in Florida occupy habitats that do not exist in Florida. These are also illustrated here for collectors in northern states, and to illustrate habitat preferences for all eastern species.

Roadside

Roadsides are subject to much disruption, and generally harbor an assortment of grasses and weeds that are undesirable for eastern species of tiger beetles (plates 9–10). A single species, *C. punctulata*, may be found here with consistency. When a road intersects a particular soil type, and erosion creates exposed banks, other species may be found; however, roadside habitat is a poor habitat when it comes to tiger beetles.

Scrub

Scrub habitat (plates 11–12) contains a variety of species of pine and oak, with a drained, sandy substrate. Typically this habitat is referred to as "sand-pine scrub" or "turkey oak scrub" or just "oak scrub." This habitat is the preferred habitat of several species of eastern tiger beetles, in particular *scutellaris* (several varieties), *abdominalis*, *gratiosa*, *hirtilabris*, *nigrior*, *scabrosa*, and *highlandensis*.

Rivers, stream banks, and sandbars (riparian)

Many eastern species of tiger beetles occupy habitat adjacent to rivers or on the margins of ponds and lakes (plates 13–18). Northern species that share this habitat include *repanda*, *ancocisconensis*, *duodecimguttata*, *marginipennis*, and *puritana*,

Southern riparian species include *lepida, wapleri, blanda, cursitans, hirticollis, macra, cuprascens, formosa generosa, trifasciata,* and *Megacephala carolina.*

Seashores and coastal dunes (littoral)

Much of the eastern seashore habitat in Florida has been destroyed by development and vehicular traffic. Where suitable habitat exists, *dorsalis media, hirticollis,* and *marginata* may be found; on the west coast, *dorsalis saulcyi,* and *hamata lacerata* make their home along the shore (plates 19–20).

Coral outcropping

One species, *olivacea,* occupies this habitat in the outermost Keys of Florida (plate 21). Much of the habitat has been destroyed by development, and recent trips to locate this species have failed. Occasionally *hamata lacerata* and *marginata* occur in the same habitat.

Gravel pits and sandpits

Gravel pits are a more northern phenomenon (plates 22–24). Where they occur, *formosa generosa, duodecimguttata, tranquebarica, longilabris,* and *scutellaris lecontei* may be found.

Woodland paths and roads

Any open woodland path or road may harbor tiger beetles (plates 25–26). Northern species may be *purpurea, sexguttata,* and *patruela,* while southern species may include *splendida, unipunctata, rufiventris, sexguttata,* and *viridicollis* (Cuba).

Eroded clay banks

Northern species found on eroded clay banks include *limbalis,* while southern species may include *tranquebarica, rufiventris, sexguttata,* and *splendida.*

Salt marshes

Along the eastern coast, *marginata* and *striga* are found in salt marshes from South Carolina to Florida, and *hamata lacerata* and *striga* are found along the west coast of Florida in this habitat (plates 27–28).

Coastal alkali mudflats

This habitat contains the largest number of species in Florida. Depending on locale, *C. striga, severa, togata, marginata, hamata lacerata, trifasciata ascendens, Megacephala carolina floridana,* and *M. virginica* may occur together.

Rock outcropping

A single northern species in Massachusetts, *rufiventris hentzii*, occupies this habitat exclusively (plate 29). Occasional specimens of *sexguttata* may occur here also.

Plate 9. Eglin AFB, Florida. Roadside habitat of *C. nigrior* and *punctulata.*

Plate 10. Route 24, Levy County, Florida. Roadside cutting through oak scrub–sandhills, Cedar Key scrub, habitat of *C. scutellaris unicolor, hirtilabris,* and *scabrosa.*

Plate 11. Josephine Creek, Highlands County, Florida. Scrub habitat, type locality for *C. highlandensis.*

Plate 12. Two miles west of Newberry, Florida. Turkey oak scrub habitat of *C. abdominalis, hirtilabris,* and *scutellaris unicolor.*

Plate 13. Apalachicola River, Torreya State Park, Liberty County, Florida. Habitat of *C. blanda* and *repanda*.

Plate 14. Blackwater River, Florida. Sandbar habitat of *C. wapleri, blanda, repanda,* and *Megacephala carolina*.

Plate 15. Escambia River, Florida. Shoreline habitat of *C. wapleri, repanda, hirticollis,* and *Megacephala carolina*.

18 · Florida and Eastern U.S. Tiger Beetles

Plate 16. Arkansas River, Pine Bluff, Arkansas. Shoreline sandbar habitat of *C. hirticollis*, *cuprascens*, and *formosa generosa*.

Plate 17. Connecticut River, Walpole, Vermont. Habitat of *C. repanda* and *marginipennis*.

Plate 18. Gulf coast, Alligator Point, Wakulla County, Florida. Habitat of *C. dorsalis saulcyi* and *C. hamata lacerata*.

Plate 19. Little Talbot State Park, Duval County, Florida. Habitat of *C. dorsalis media*, whose larval burrows are indicated by red flags, and *marginata*, which dwell in the mudflats.

Plate 20. Little Talbot State Park, Duval County, Florida. Habitat of *C. dorsalis media*.

Plate 21. Grassy Key, Monroe County, Florida. Gulf-side coral outcropping habitat of *C. olivacea*, on coral, and *C. hamata lacerata*, on sand and mud.

Plate 22. New Hampshire. Habitat of *C. formosa generosa.*

Plate 23. Maine. Overgrown gravel pit, habitat of *C. longilabris.*

Plate 24. New Hampshire. Habitat of *C. tranquebarica* and *duodecimguttata.*

Plate 25. (*above left*) Unicoi State Park, White County, Georgia. Habitat of *C. splendida, sexguttata, unipunctata,* and *rufiventris.*

Plate 26. (*above right*) Clayton, Georgia. Power line cut. Same species found here as in Plate 25.

Plate 27. Shell Mound, Levy County, Florida. Tidal salt marsh, habitat of *C. hamata lacerata.*

Plate 28. South of Steinhatchee, two miles north of Horseshoe Beach, "Road to Nowhere," Route 361, Dixie County, Florida. Habitat of *C. severa, togata, hamata lacerata, marginata, striga, trifasciata ascendens, Megacephala virginica,* and *M. carolina floridana.*

Plate 29. Blue Hill State Park, Suffolk County, Massachusetts. Habitat of *C. rufiventris hentzii.*

Habitat segregation within sympatric populations of tiger beetles

When more than one species of tiger beetle appear to coexist in the same habitat (sympatry), closer inspection will usually reveal subtle differences in microhabitats. For example, at a coastal mudflat in South Carolina (plate 33), three species—*marginata* (plate 30), *dorsalis media* (plate

31), and *hirticollis* (plate 32)—occur more or less together. Close examination inspection reveals that *marginata* is most abundant on the wettest part of the mudflat, *dorsalis media* occupies a middle ground where the soil is moist but not muddy, and *hirticollis* is most abundant where the soil is most sandy but still moist. Individuals of *dorsalis media* also occur on the beach, but in fewer numbers. The other two species are absent on the main beach.

 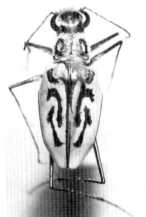 

Plate 30. *C. marginata.*   Plate 31. *C. dorsalis media.*   Plate 32. *C. hirticollis.*

Plate 33. Huntington Beach, Georgetown County, South Carolina. Marsh area behind beach dunes. Within this habitat three tiger beetle species occur, each with its own preferred section of mudflat. *Cicindela marginata* is most abundant in the wetter, muddy area adjacent to water. *C. dorsalis media* occurs throughout, but is most abundant in the mid-area of the mudflat. *C. hirticollis* prefers the higher, drier section of the mudflat closest to beach sand dunes.

3

Florida's Geological History and Its Relationship to Animal and Plant Distributions

When studying the distributions of animals or plants one must take into consideration the geological history of the region of interest. Knisley and Schultz (1997) discuss the geological history of eastern United States and its impact on the evolution and distribution of tiger beetle species, but do not mention Florida. Florida is unique in many ways and differs from the other eastern states. I discuss here several theories of the geological history of the state and its apparent impact on the distribution of plants and animals.

If you are unfamiliar with geological time period names, refer to table 1 at the end of this chapter. Florida's geological history is complex and still debated. Early studies suggested that the state was submerged until the Lower Oligocene (forty million years ago), when a large central island was believed to have existed (Schuchert 1929). During the Miocene (twenty-five million years ago) Florida was reconnected with the mainland, remaining connected through the Pliocene (eleven million years ago). Various parts of the Pleistocene saw Florida broken into many islands, later to be reconnected with the mainland. This insular development of Florida, as well as the high degree of endemism and disjunction of peninsular fauna from the mainland has resulted in many and varied explanations. Many systematic treatments rely on the Pleistocene sea levels, the breakup of Florida into islands, and the reconnection to the mainland as explanation for species found peculiar to Florida.

Olsen, Hubbell, and Howden (1954), in discussing Florida and the scarab genus *Mycotrupes*, state, "Allopatric distribution of closely related but distinct species causes no surprise when it is encountered in archipelagoes or mountain areas, where the geographic or ecological isolation is apparent . . . To find a region where, in small space and relatively uniform environment, many animal groups are represented by a number of very distinct species with mutually exclusive ranges, is a phenomenon that calls for explanation, especially when in this region similar or complementary distributional patterns tend to occur in unrelated species groups."

Prior to new geological evidence concerning the age of Florida, taxonomists were left with the Pleistocene events to explain endemism in the peninsula. However, the suggested speciation processes required a much longer existence, a pre-Pleistocene existence in Florida. This dichotomy of evidence versus needed processes led to dilemmas such as the following (Olsen, Hubbell, and Howden 1954):

> The wide morphological divergence of *Mycotrupes* from other Geotrupini connotes a long independent evolutionary history, and in this connection with the distributional evidence below leads me to conclude that the genus arose a long time ago in the region which it now occupies and that it was already differentiated and in residence by middle Pliocene times . . . My third assumption is an act of faith, based on grounds that seem good to me and most zoogeographers, but not to geologists. It is that there has been land in central Florida since the Pliocene, at least in the form of a few reasonable-size islands . . . The case for *gaigei* is the weakest, in that it hypothecates a group of persistent central Florida islands that perhaps did not exist . . . It would be interesting to see the methods here used applied to other better known groups of Coastal Plain animals.

Apparently unnoticed, Applin and Applin (1944) had stated that "The area in northeastern Florida and the central part of the peninsula where Oligocene beds are absent probably existed as an island (Orange Island) in the Oligocene sea, separated from the mainland on the northwest by a channel (Suwannee Strait) approximately 100 miles wide which extended southwestward across Georgia through the Tallahassee area of Florida to the Gulf of Mexico." Applin and Applin (1944) show an island in Upper Eocene encompassing parts of Polk, Lake, Pasco, Hernando, Citrus, Sumter, Marion, Alachua, Gilchrist, Dixie, Columbia, Suwannee, Madison, Taylor, and Lafayette counties. Orange Island (Oligocene) included parts of Martin, Glades, Highlands, Okeechobee, St. Lucie, Indian River, Osceola, Polk, Brevard, Orange, Lake, Sumter, Citrus, Volusia, Marion, Levy, Flagler, Putnam, Alachua, Dixie, Gilchrist, Lafayette, Suwannee, Columbia, Union, Bradford, Clay, St. Johns, Duval, Nassau, and Baker counties. Here was the large island wished for by Hubbell. Hubbell (1960) was finally able to state, "Even more important is the discovery of good geologic evidence that a rather large land mass has been in continuous existence in north Florida since the middle Miocene (Vernon 1951)." Apparently Vernon was equally unaware of the work by Applin and Applin (1944), where there is also a statement of the relationship between Florida, the Antilles, and Mexico: "the fauna of the lower Cretaceous limestones of south Florida again resembles that found in certain derived deposits in Cuba and also that of the El Abra limestone of southern Mexico (Cenomanian-Albian) there designated as Middle Cretaceous." Here is one of the first suggestions of a time and faunal relationship between Florida and Mexico as early as the Cretaceous.

Geological evidence indicated that, in fact, Florida had land much older than previously known by geologists, but suspected by zoogeographers. That Florida has been separated from the mainland is fact. That islands existed during various glacial eras is also fact. The interpretation of these facts introduces the debate concerning the derivation of Florida's fauna and flora. According to Croizat (1958, 1962), the repeated occurrence (track) of distribution in unrelated groups should be statistically significant and worthy of generalized statement. The following examples are cited for two reasons: to document repeated distribution patterns, and to illustrate various attempts at reconciling data. These examples are the result of months of literature search. I am sure there are others.

Speaking of the insect order Lepidoptera, Evans (1955) states, "As would be expected, if butterflies were first evolving prior to continental separations by seafloor spreading, the primitive Hesperiidae display the greatest number of links between African and South American butterfly faunas. Evans (1949, 1951, 1952) points out these. In Pyrginae, *Celaenorrhinus* is the only tropical genus occurring unchanged in America, Africa, and the Oriental region. . . . The genera *Talides* of America, *Gamia* of Africa, and *Gangara* of Asia resemble one another closely enough to indicate a common ancestor."

Discussing North American salamanders, Lowe (1950) states,

> Thus the present latitudinal disjunctive *Aneides* falls within a fundamental pattern common to many plants and animals groups, a pattern in accord with paleontological and meteorological facts

and inferences therefore to past climates throughout the Cenozoic. If we can regard the disjunction in range as no younger than mid-Tertiary, as seems indicated by the distribution and history of the forest type with which they now occur, then it seems probable that the species of *Aneides* . . . have been diverging slowly through the last 20–30 million years in earth's history, and have not arisen during the Pleistocene.

Previous authors had attempted to explain the speciation of *Aneides* strictly by Pleistocene events.

If Florida did exist as a large island, when did this island exist and what was the extent of its land mass? Schuchert (1929) shows that Florida was completely submerged as late as the Upper Eocene. Applin and Applin (1944) show a large island already in existence in the Upper Eocene. Schuchert (1929) believed that it was during the Oligocene that an island emerged from the ocean. St. John (1936), while discussing the ferns of central Florida, and the geological history of the area, states,

> Considering the general climatic conditions at that period . . . and that the tropical sea flowed north of this island, it seems certain that its flora was wholly tropical. . . . Prof. Schuchert believes that this island, varying in size at different periods, persisted for at least 12,000,000 years before connection was made with the mainland. This long period of isolation, with changes in climate and other ecological factors, accounts for the development of the endemic species. . . . At about the same geological time (Miocene) the emergence of the northern half of the peninsula connected it with the mainland, making northern migration possible while tropical conditions still prevailed in the southern part of the continent. . . . During the Pleistocene the successive glacial periods sent waves of colder climate southward, introducing some strictly northern species. . . . The tropical ferns that are now isolated in central Florida are the wreckage of the once far richer flora of the Oligocene island . . . with some apparent exceptions to the rule. The present distribution of two species *Ophioglossum engelmanni* and *O. crotalophoroides*, seems to indicate that they reached Florida and the adjacent states at the north by way of Mexico.

Here is illustrated the first of several distributions that when put together produce a general track (*sensu* Croizat 1962) that I believe can more satisfactorily explain these phenomena. St. John (1936) required a succession of waves of migrations, both from the north and the south, as well as from Mexico, to explain the localized ferns in central Florida. If the flora of this Oligocene island that emerged from a sea were wholly tropical, where did it come from? According to Schuchert (1929), the rest of the Gulf of Mexico was under water at this time, and even if there were islands present, did this mean that an entire flora had to suddenly appear on this island? An entire fauna? If we carry this sequence of events backward in time, where did these organisms come from if no land existed? I would suggest now, and will suggest after more examples, that Florida has never been completely submerged since the Cretaceous, and the fauna and flora that are recorded from the Oligocene are remnant of a much wider spread biota of much older age. Therefore, when Florida became an island during the Eocene and Oligocene, it already had a fauna and flora with affinities that will be discussed later.

Where did the fauna and flora come from, the organisms that we see remnants of in Florida, the relicts and endemics? Mice have a representative in Florida that lends some clues to these questions. Johnson and Layne (1961) discuss the mouse genus *Peromyscus* in Florida:

> *Peromyscus floridanus* is a distinctive species of the wide ranging and varied genus *Peromyscus*. . . . This species has also been recorded from the Pleistocene in Florida (Sherman 1952). . . . A recent study by Hooper indicates that . . . *P. floridanus* most closely resembles *P. lophurus* and *lepturus* from southern Mexico and Guatemala. . . . Presumably a once continuously distributed, warmth adapted coastal plain form of *Peromyscus* was forced into separate refuges in Florida and southern Mexico or Central America during glacial advances, the isolated elements of the original population subsequently undergoing evolutionary divergence in the widely separated eastern and western centers. . . . The logical explanation of this particular host-flea relationship seems to be that ancestral *Peromyscus* stock was infested with a species of *Polygenis* whose range fragmented along with its host's during the Pleistocene.

Here we have a mouse genus and its flea parasite, both of which have diverged from ancestral forms known to have come from southern Mexico or Central America. If the mouse *P. floridanus* is known from

Pleistocene fossils in Florida, doesn't that mean it was probably in existence before the Pleistocene? Such may have been the case, as with *Aneides* (Lowe 1950). Blair (1950), in discussing two other species of *Peromyscus,* explains that the five-hundred-mile disjunction between *P. polionotus* and *P. maniculatus pallescens* can be explained if "we assume continuous distribution of *maniculatus* across the coastal plain in Pleistocene, with encroachment of the Gulf during an interglacial, part of the population became isolated in Florida. The coastal plain population of *maniculatus* disappeared eastward of Texas, effectively isolating the Florida population. An ecological barrier now separates the two coastal plain populations, for the forests and coastal marshlands now occupying the coastal plain eastward from eastern Texas are avoided by both species."

Geological evidence to the contrary, here is an explanation that fails to take into consideration available data. By most authorities, a continuous range from southern Mexico to Florida during the Pleistocene would have been impossible. Such a range could have existed only during the Miocene or even during the Oligocene or Eocene. Therefore, Blair (1950) has failed to take into account the available albeit controversial geological information available. The ranges of these mice or their ancestors were probably disrupted during the Eocene, when Florida was a large island.

Hubbell (1960) used migration from a Mexican stock into Florida to explain speciation within the orthopteran *Paratettix:*

> Lest the hypothesis of long-distance migration of southwestern and Mexican stocks into the Florida land mass seem speculative, one item of fossil evidence can be cited to show that such movements of Orthoptera did occur. In the Pamlico sands of late Pleistocene age, at Vero, Florida, numerous well preserved specimens of tetrigid locusts have recently been found . . . most belong to the species *Paratettix rugosus,* today abundant in, but restricted to, Florida and south Georgia. A number, however, are assignable without question to *Paratettix toltecus,* a Mexican species which now occurs no closer to Florida than southern coastal Texas, but which is, on morphological evidence, the presumptive ancestor of *rugosus*. From this one may deduce that at some time, probably pre-Pleistocene or early Pliocene, ancestral *toltecus* spread east into Florida; that its descendants isolated in that region, evolved into *rugosus;*

and that during the late Pleistocene *toltecus* again reached Florida and then lived for a time with its offspring *rugosus* before dying out in that region.

The occurrence of these two related species in the same strata suggests that more needs to be known about them. Both *toltecus* and *rugosus* are extant species. Their presence together in the same series precludes the assumption that they are closely related, because sympatry is believed to be more prevalent among more distantly related taxa. If the statement "sympatry . . . is itself evidence for dispersal" is accepted, then *toltecus* is not the close relative of *rugosus* as was then believed (Croizat, Rosen, and Nelson 1974).

Termites are not excluded from this Central America/Mexico relation with Florida. Emerson (1952: 219) discusses the relationship of several genera of termites. Again the same relation is suggested:

> *Calcaritermes* is directly derived from *Glyptotermes* and is closely related in both its morphology and ecology. With the exception of a single species in north and central Florida, this genus is wholly Neotropical, with the center of dispersal in Central America. . . . Because of its limited distribution, it may be postulated that the genus arose in the New World, probably in Central America, during Tertiary times. . . .
>
> *Prorhinotermes* is primarily confined to islands and is absent on all continents except on the coast of Central America and southern Florida (Allee et al. 1949: 725). It is found in the West Indies, Pacific Islands, East Indies, and islands of the Indian Ocean including Madagascar. Contrary to most groups of termites, this genus seems to be distributed in floating logs. It survives on islands where competition is weak, but is eliminated on the continents where these termites must often be washed ashore. The center of dispersal seems to be the islands of the Pacific . . . and the genus probably reached the West Indies at the time of sea connections through Panama during Tertiary times.

Spiders are not excluded either from the use of Pleistocene events to explain their distributions. McCrone and Levi (1964: 25) discuss species of the black widow genus *Latrodectus*:

> *Latrodectus bishopi* shows a very distinct habitat preference. It is completely restricted to inland, dunelike areas that support a plant

association called sand-pine scrub. The vegetation is xeromorphic and is dominated by the sand pine (*Pinus clausa*). . . . It is known that the sand-pine scrubs inhabited by *L. bishopi* arose in connection with islands that were present in the Florida area during the Pleistocene (Laessle, 1958). The literature (Neill, 1957) records a number of species that are endemic to these scrubs or other south-central habitats in Florida. Many of these species are closely related to other species whose ranges extend only into northern Florida. Thus the conjecture can be made that *L. bishopi* differentiated from *L. variolus* on an isolated island or island group in the Pleistocene seas.

Brady (1972: 31–33) uses a similar argument in discussing the genus *Sossipus:*

> The distribution patterns of different populations of *Sossipus* in Florida appear to reflect recent geologic events. *Sossipus placidus* is restricted to an area represented by Red Hill Island of the Aftonian Interglacial (Laessle, 1958). . . . The rise in sea-level during the Pleistocene with the concomitant production of various islands where the Florida peninsula now stands has influenced speciation in at least three different genera of spiders. . . . I suggest that the reduction of population numbers on the Pleistocene islands with a corresponding reduction in genetic variability played a significant role in the process of speciation. These geographical isolates became genetically homogeneous and ecologically specialized for the xerophytic conditions of the Pleistocene islands. Even today the limits of the species ranges of these spiders is prescribed by the extent of the xeric communities formed from these islands. When the islands were joined after the Pleistocene, the insular populations were effectively isolated ecologically and reproductively.

If the recurring phenomena of insects appearing to have their nearest relatives in Florida and Mexico is not unusual enough, the birds also have at least one representative showing the same. The scrub jay *Aphelocoma coerulescens* is limited to peninsular Florida, found in scrub that is also shared by *Cicindela scabrosa* or *C. abdominalis*. Pitelka (1951) characterized the habitat as "a low, dense, semixeric thicket usually not exceeding 10 feet in height and commonly dominated by oaks (*Quercus geminata, Q. myrtifolia*), sword palmetto (*Serenoa serrulata*), dwarf wax myrtle (*Myrica pumila*), and scattered stunted pines (*Pinus clausa* or *P. caribea*)." These are the trees found with *C. scabrosa,* and the Caribbean pine is restricted to the south part of Florida in isolated scrubs.

The scrub jay has as its nearest relative a southwestern form and a Central American form. Pitelka (1951: 383–384, 386) explains the distribution of these jays:

> It is maintained by students of paleobotany that the North American vegetation attained essentially its present aspect and relative distribution by the end of the Pliocene. . . . The sclerophyll woodland with which *A. coerulescens* and *A. ultramarina* are associated spread during the Miocene, reaching southern and central California, the Columbia plateau, and the plains region during the Miocene, and apparently became most extensive in the late Miocene and early Pliocene. Probably at this time, sclerophyll woodland extended eastward along the Gulf Coast. The extension apparently existed along the southern and eastern borders of the southern Appalachian upland over a belt that has been exposed since the beginning of the Tertiary (Fernald 1931: 27). The pines and oaks of the southern states which are related to the sclerophyll woodland of Mexico apparently are derivatives of vegetation having a history in that region which dates from at least the early Pliocene. Exposure of the Florida peninsula occurred during the latter part of the Tertiary and the Quaternary. The presence of certain species of plants and animals in Florida which are isolated on the peninsula may be explained by the early eastward extension of the sclerophyll woodland. Those species represent, in part at least, segregates or relicts the separation of which from ancestral stocks is related to regional biotic shifts or fluctuations probably imposed by successive Pleistocene glaciations and the consequent southward invasions of boreal forests (Brown 1938; Davis 1946; Potzger and Thorp 1947). Few concrete examples of these segregates are available, since little effort has yet been made to study the relationships of the Florida biota to that of western North America. *A. coerulescens* is a unique example among scrub-inhabiting birds. Among plants, *Ceanothus* (Mason 1942: 290) and *Eriogonum* and *Lygodesmia* (G. L. Stebbins, oral communication) indicate derivation of

the Floridian forms from the southwest. Other good examples undoubtedly exist in different animal groups as well as plants. . . . The dispersal of *A. coerulescens* in North America is apparently related to the spread of Mexican sclerophyll woodland and chapparal assemblages into southwestern North America and the Gulf and South Atlantic coastal regions in the early Pliocene.

If the examples of Florida forms being derived from southwestern migrations are numerous, examples also exist for the reverse. Blair (1958: 453) discusses species of mice in the genus *Peromyscus:* "*Peromyscus gossypinus* of the coastal plain and *P. leucopus* overlap in a generally narrow zone along the border of the coastal plain and along the deciduous forest border in eastern Texas and Oklahoma (Osgood 1909; McCarley 1954). The present distribution of this species pair can be explained as a result of the westward and northward spread of the coastal plain adapted *gossypinus* from a refuge in Florida and the northward and eastward spread of *leucopus* from a Mexican refuge, where it could have been adapted to less mesic upland forests."

If those migrations occurred east to Florida from the southwest or Mexico, and west from Florida as suggested above, what stimulus for such massive migrations might have taken place? Blair (1958) states that the present distribution of vertebrates in southern United States, on the Gulf and coastal plains, can be explained only by the hypothesis of drastic ecological changes in the Pleistocene: "The argument is essentially that at peaks of glacial advances into the northern United States climatic and ecological changes in the southern United States were so great as to drive warmth-adapted species into separate refuges in Florida and Mexico." Braun (1955), however, argues that the conditions that produced the Pleistocene glaciations had little effect in the southern United States.

The flora of the coastal region (Austroriparian) extends with only minor variation from eastern Texas to the Atlantic Coast (Blair, 1958). Southern grasslands border these forests to the west and north. These grasslands are of late Miocene age, with great development in the Pliocene (Clements and Chaney 1937), and are a barrier to the westward distribution of many forest animals. Do not the forests then become barriers to the eastward migration of grassland animals? Such spread of biotas must then have taken place before the Miocene!

Intent upon explaining distributions with the Pleistocene, Blair (1958:

445) lists numerous examples of disjunctions between the southeast and southwest:

> The Floridian fauna also includes a considerable element of species that belong to groups with centers of distribution in the Southwest, where they are generally adapted to more xeric conditions than exist today on the coastal plain. The presence of this western element in Florida suggests past climatic fluctuations on the coastal plain that favored eastward spread.... Some of these (e.g., *Microhyla carolinensis* and *Peromyscus gossypinus*) have become adapted to high-moisture situations of the coastal plain. Others have retained their xeric adaptations and exist today in the most xeric situation available. The most extreme examples of the latter group include *Scaphinopus*, *Cnemidophorus*, *Aphelocoma*, and *Peromyscus polionotus*.

In discussing tree distribution Blair (1958: 452) states:

> An area on the floodplain of the San Marcos River in Gonzales County, Texas, has an assemblage of eastern coastal plain plants and animals that are disjunct from their main populations. The palmetto (*Sabal minor*), burr oak (*Quercus macrocarpa*), wax myrtle (*Myrica cerifera*), and ash (*Fraxinus*) are representatives of a rather large number of plants in this category. The vertebrates include the canebrake rattlesnake (*Crotalus horridus*), banded watersnake (*Natrix sipedon*), and narrow-mouth frog (*Microhyla carolinensis*).
>
> ...
>
> Both forest and grassland occur in the hiatus between the eastern and western populations of the indigo snake (*Drymarchon corais*). The eastern population occurs on the coastal plain east of the Mississippi Embayment; the western population ranges from the area of Corpus Christi in Texas southward into northern South America. (Blair 1958: 455)
>
> ...
>
> A very wide disjunction exists between an eastern treefrog, *Hyla femoralis*, and the apparently related *H. arenicolor* of the west. *H. femoralis* occurs on the coastal plain west to the embayment and *H. arenicolor* ranges westward from trans-Pecos Texas, which

means there is a gap of some 700 miles between the ranges. (Blair 1958: 457)

. . .

The *Peromyscus maniculatus* group of mice, with one of the most complex distributional patterns of any North American vertebrate . . . shows east-west speciation on the coastal plain (Blair, 1950). The beach mouse (*P. polionotus*), which occurs on the coastal plain east of Mobile Bay, on morphological evidence is derived from the grassland-adapted ecotype of the deermouse (*P. maniculatus*), which today ranges southward into south-central Texas. The beach mouse presumably originated through an eastward dispersal along Gulf Coast beaches and subsequent isolation in Florida. (Blair 1958: 459)

. . .

The distributional patterns of warmth adapted vertebrates on the coastal plain as discussed above are overwhelmingly indicative of east-west fragmentation of ranges as the initial agent of geographic speciation in this fauna. (Blair 1958: 461)

. . .

Sympatric distributions of coastal plain groups may trace back to the same kind of east-west disjunction exhibited by the allopatric populations, but such history is difficult to demonstrate. McConker (1954) has theorized that the three species of legless lizards (*Ophisaurus*) of the coastal plain originated through east-west splitting in the Third Glacial (Illinoian) and through subsequent isolation of the third population on Florida islands in the Third Interglacial (Sangamon). (Blair 1958: 462)

Numerous other examples are found in the literature to illustrate similar explanations. Howden (1961, 1963, 1966, 1969) cites many examples, and gives a good summary of the thought concerning the effect of the Pleistocene on North American insects.

Having cited these many examples, how does this apply to Florida tiger beetles?

There are only a few species in Florida that do not occur elsewhere in eastern United States, or at least in southeastern United States. Two species occur only in Florida and Cuba (*Cicindela olivacea* and *viridicollis*).

Species restricted to Florida are *C. hirtilabris* (found only in Florida and extreme southeast Georgia), *C. scabrosa* (Florida only), and *C. highlandensis* (central Florida only). These restricted ranges reflect isolation for periods of time such that new species had time to evolve while Florida was a series of smaller islands separated from the mainland. When a connection was reestablished these forms were sufficiently different that they could no longer interbreed with mainland populations of their sister species.

Table 1. Geological timescale

Period	Epoch	Years to present
Quaternary	Pleistocene to present	2 million
Tertiary	Pliocene	11 million
Cenozoic		
	Miocene	25 million
	Oligocene	40 million
	Eocene	60 million
	Paleocene	70 million
Mesozoic		
	Cretaceous	135 million
	Jurassic	180 million
	Triassic	225 million
Paleozoic		
	Permian	270 million
	Pennsylvanian	310 million
	Mississippian	350 million
	Devonian	400 million
	Silurian	440 million
	Ordovician	500 million
	Cambrian	600 million

4

Species Criteria

Each tiger beetle discussed here is recognized by its Latin binomen (two names) or in a few cases trinomen (three names). The first name, capitalized, is the genus; the second name the species, and in some cases a third name is the variety or subspecies. The naming authority is placed at the end of the scientific name. The entire scientific name applied to an insect constitutes its species name. But what constitutes a species? A subspecies? Within each group of organisms the answer may vary according to the authority one consults and the existing knowledge of that group. Many opinions have been offered about the problems surrounding species definitions, but Hubbs (1943: 120–121) perhaps summed it up best:

> To my knowledge no single criterion that has ever been erected will suffice to define the species, without the need for some exceptions and modifications. The more intensely species are studied, throughout their ranges, the more difficult it often becomes to decide on the taxonomic rankings. . . . But neither in detailed taxonomic treatment nor in general speciational theory should we forget the true situation. Arbitrary decisions must often be made, to meet the demands of the Linnean system of zoological nomenclature, but it is bad science to deny that the decisions are arbitrary. Neither conventional views nor subjective subterfuges—whether by the old-line systematists or by the modern speciationist—can transcend the facts, or create a simple "correct" system of taxonomy or a simple theory of speciation out of a situation that is inherently complex. Evolution has been and remains at work.

Most taxonomists have a feel for species within their specialty, but their criteria are far from standardized. Reproductive isolation, resulting in evolution of genetic incompatibility (sterile hybrids), has often been proposed as the ideal requirement. But this genetic incompatibility is very difficult, if not impossible to demonstrate. Therefore, taxonomists infer reproductive isolation from several demonstrable characters, including geographical separation (allopatry), seasonal disjunction (allochrony), ecological separation, morphological differences (especially genitalic differences), behavioral differences, and host preferences. Any one of these in and of itself may not be an indicator of reproductive isolation; however, one or more may make a strong case for species difference.

Ecological separation (habitat difference) is perhaps one of the strongest indicators of species differences. In spite of lack of obvious morphological variations, insects occurring in distinct habitats should be closely investigated.

Allochronic speciation, the result of adaptations to different climates and seasons, probably occurs more often than taxonomists realize. Distributions that resulted from such geological events as the Pleistocene ice ages must have produced allochrony in some insects. This is evidenced by insects in Florida that are active as adults while their northern counterparts are either immature or buried under snow and ice.

Genitalic characters are often treated as the ultimate criterion for species separation. I believe this to be simply one more character that by itself should not be weighted so heavily. Lack of genitalic differences should not rule out species differences. Closely related species would be expected to show similar genitalic characters.

Museum collections lack most ecological information. Chronological data may reflect seasonal abundance of collectors, not insects, and should be carefully considered before suggesting allochrony. Geographic separation may again be a function of collections, and may not reflect the natural ranges of a particular form. I feel that a species should not be defined on purely morphological terms, but realistically this may be the only means available.

Even more unsettling is the subspecies category. Within the genus *Cicindela* subspecific names have been widely applied, often without adequate study of the overall variation within the particular species in question. For this reason, the taxonomic validity of many of the proposed subspecific names is uncertain and cannot be resolved without detailed analysis of the forms involved.

Indeed, the validity and usefulness of subspecific names have been debated for years. Wilson and Brown (1953) present a resume for the arguments against the application of subspecific names. Different opinions on this topic are expressed in Brown and Wilson (1953), Edwards

(1954), Fox (1955), Gillham (1956), Hubbell (1954, 1956), and Rogers (1955). If subspecific names are used, I believe that definite criteria should be met with consistency. Within tiger beetles the universally applied subspecies criterion has been color and degree of maculation, with too little attempt at analyzing other characters as well. Exceptions to this may be found in Freitag (1965, 1979), Rivalier (1950a,b, 1953a,b), Willis (1967), and Rumpp (1956, 1957, 1961).

The subspecies concept is basically geographical, including the probability for intergrades in characters at zones of contact. Lidicker (1962: 169) defined a subspecies as "a relatively homogeneous and genetically distinct portion of a species which represents a separately evolving, or recently evolved, lineage with its own evolutionary tendencies, inhabits a definite geographical area, is usually at least partially isolated, and may intergrade gradually, although over a fairly narrow zone, with adjacent subspecies."

If subspecific names are applied to differing populations without geographical definition and the potential for interbreeding, two serious risks are incurred: (1) the naming of a form that is merely an expression of the variability within a species range, and (2), more important from the systematic standpoint, obscuring recognition of sibling species. The former problem can usually be resolved by simply stating that two subspecies should not be sympatric (sharing the same range) except at the limits of their ranges. Within this zone of overlap, intergrades should be expected to occur.

The problem concerning sibling species (species that do not appear to differ in external morphology but behave as distinct species) involves the current trend in systematics to define species on the basis of biological rather than typological criteria. The morphological basis for species separations has proven inadequate, and biological criteria are used to further define species. Mayr (1974) discusses this matter in greater detail. As early as Charles Darwin this problem was recognized: in *The Origin of Species,* Darwin ([1865] 1968: 455–456) states,

> Hereafter, we shall be compelled to acknowledge that the only distinction between species and well marked varieties is, that the latter are known, or believed, to be connected at the present day by intermediate gradations, whereas species were formerly thus connected. . . . In short, we shall have to treat species in the same manner as those naturalists treat genera, who admit that genera are merely

artificial combinations made for convenience. This may not be a cheering prospect, but we shall at least be freed from the vain search for the undiscoverable essence of the term species.

It is sufficient to state that the biological species may be defined as a multidimensional concept, involving morphology, ecology, behavior, and additional criteria that may be found to support or correlate with the reproductive integrity of the organism in question. In my opinion, a subspecies should be recognized only if the geographical criterion with a defined zone of intergradation is met. A species is defined as a group of reproductively isolated organisms, differing in one or more characters (other than color) unique to the entire range of the organisms involved. Allopatric (separated geographically) populations constantly differing in at least one character are considered specifically distinct, and sympatric (sharing the same range) forms lacking intergrades are also considered separate species. Our classification system is admittedly artificial. Schmidt (1950: 333) summed up the problems and potential of taxonomic classification:

> The system of nomenclature, as formulated under an elaborate set of international rules, lends itself to being played as a game, with personal and group (museum) rivalries, and personal and group vanities as the motivations instead of the advancement of science. Thus in the past most species and subspecies have been described without any adequate knowledge of what was being described; and the rush to describe and name makes it necessary to redescribe and recombine when more adequate knowledge accumulates.... When systematists engage in sound studies, in which accurately mapped ranges replace the guesswork of early descriptions of species, and in which the range of variation of the component populations of the species begins to be understood, systematics may again take its rightful place among the biological sciences as the essential background for ecological and genetic and evolutionary studies.

5

Classification and Identification of Tiger Beetles

Status of classification

Both suprageneric and generic classifications of tiger beetles are reasonably sound and most genera are readily identifiable. Species definition within each group, however, is a problem. Worldwide, there are an estimated two thousand species within the subfamily Cicindelinae (Pearson 1988), with 125 species known from North America (Pearson 1948). Within these, there are abundant subspecies as well as color variants, many named and overlapping, producing a confusing synonymy. Much work must be done before all species can be properly understood; however, many species groups and regional fauna have been studied so that now it is possible to place names on most forms found in North America. There remain several named forms of *Megacephala* to be interpreted, and the genus *Omus* contains either a few or many species, depending on the interpretation of its many allopatric populations. Several new species of tiger beetles have recently been described (Choate 1984; Freitag, Kavanaugh, and Morgan 1993; Johnson 1991a,b, 1994; Mateu 1975; Rumpp 1967, 1986).

Willis (1969) provides a translation of Horn's (1868) key to the world genera of tiger beetles, and Reichardt (1977) provides keys to the tribes and genera of Neotropical Cicindelinae. Larochelle (1986d) lists references relevant to distribution of Cicindelidae of North America north of Mexico.

For determination to species of regional faunas see the following: Acciavatti, Allen, and Stuart (1992) (West Virginia), Bertholf (1983) (Arizona), Boyd (1978) (New Jersey), Carter (1989) (Nebraska), Cartwright (1935) (South Carolina), Cazier (1948) (lower California), Cazier (1954) (Mexico), Ciegler (1997) (South Carolina), Dawson and Horn (1928) (Minnesota), Downie and Arnett (1996) (northeastern United States), Drew and van Cleave (1962) (Oklahoma), Dunn (1981) (New Hampshire), Glaser (1984) (Maryland), Graves (1963 [Michigan], 1965

[Ontario]), Graves and Pearson (1973) (Arkansas, Louisiana, Mississippi), Hilchie (1985) (Alberta), Ivie (1983) (Virgin Islands), King (1988) (Mississippi), Kippenhan (1990a,b, 1994) (Colorado), Knisley, Brzoska, and Schrock (1987) (Indiana), Larson (1981) (North Dakota), Mather (1971) (Mississippi), Meserve (1936) (Nebraska), Tanner (1929) (Utah), Wallis (1961) (Canada), Wilson and Brower (1983) (Maine). Generic revisions include Willis (1968) (*Cicindela*), Vaurie (1955) (*Amblycheila*), Casey (1914a) (*Omus*), Leng (1902c) (*Megacephala* and *Cicindela*). There is no current revision of the genus *Cicindela*, but one has been in preparation for many years by M. A. Cazier.

Species groups revisions include: Acciavatti (1980) (*Cicindela praetextata*), Cazier (1937b) (*willistoni, fulgida, parowana, senilis*), Choate (1984) (*abdominalis*), Freitag (1965) (*maritima* group), Graves, Krejci, and Graves (1988) (*hirticollis*), Kaulbars and Freitag (1993c) (*sexguttata* group), Knudsen (1985) (*fulgida*), Murray (1981) (*rufiventris, sedecimpunctata, flohri*), Rumpp (1957) (*praetextata-californica*), Schincariol and Freitag (1991) (*splendida*), Sumlin (1985) (*politula*), and Varas (1927, 1928, 1929) (*formosa, purpurea, oregona*).

Key to Nearctic genera of adult Cicindelinae

1. Anterior angles of pronotum more prominent, protruding forward more than anterior margins of prosternum; anterior pronotal sulcus separated from anterior prosternal sulcus and also from prosternoepisternal sulcus (Megacephalini) 2
—Anterior pronotal angles not more prominent than prosternal margin, anterior pronotal sulcus continuous with anterior prosternal sulcus (Cicindelini) .... *Cicindela*

2(1). Posterior coxae separated; eyes small. Flightless species with fused elytra; legs and body uniform black or brown 3
—Posterior coxae contiguous; eyes large, prominent; elytra not fused, capable of flight. Legs paler than body, elytra often with pale apical lunule; upper body with green, coppery, or purple reflections ....
.... *Megacephala*

3(2). Sides of elytra widely inflexed; thorax scarcely margined; terminal maxillary palpomere shorter than third palpomere .... *Amblycheila*
—Sides of elytra narrowly inflexed; thorax distinctly margined; length of last two maxillary palpomeres subequal .... *Omus*

Classification of Nearctic genera

Some authors prefer to consider tiger beetles a separate family, divided into two subfamilies, Collyrinae and Cicindelinae. Subfamily Collyrinae is divided into two tribes, Ctenostomatini and Collyrini with two genera each, none of which are found in the United States. Subfamily Cicindelinae is divided into three tribes, Manticorini (2 genera, African only) Megacephalini, and Cicindelini with 10 and 19 genera respectively, both with representatives in North America.

Tribe Megacephalini

Amblycheila Say, 1830, 7 species, Mexico, southwestern United States, as far north as South Dakota.
Megacephala Latreille, 1802, 4 species, southeastern and southwestern United States.
Tetracha Hope, 1838. Subgenus (perhaps full genus)
Omus Eschscholtz, 1829, 4 species, with dozens of synonyms to be resolved; western United States.
Megomus Casey, 1914. synonym
Leptomus Casey, 1914. synonym

Adults of *Megacephala*, *Amblycheila*, and *Omus* are crepuscular/nocturnal and most effectively collected with pitfall traps. Some species of *Megacephala* may be attracted to electric lights, but are usually found beneath stones, hiding in crevices in the earth, or under almost any debris. Young (1980) reported *Megacephala affinis* preying on dung beetles in Panama. Guido and Fowler (1988) reported adult *Megacephala fulgida* appearing at sound traps used to attract *Scapteriscus* mole crickets. Boyd (1985) collected *M. virginica* in New Jersey using pitfall traps. Blum, Jones, and House (1981) described the chemical components of defensive secretions of *M. virginica*.

There is no current revision of *Megacephala*. Wagenaar Hummelinck (1983) provided a key to Caribbean species. New species of *Megacephala* have recently been described from Venezuela (Huber 1994, *Tetracha*), Panama (Johnson 1991b), and Nicaragua (Johnson 1993a).

Individuals within the genus *Amblycheila* have been reported as becoming active soon after sunset (Dunn 1980b). Authors frequently use the spelling *Amblychila* for the genus, but Huber (1986) pointed out that the correct spelling should be after that of Say: *Amblycheila*. Vaurie (1955) published a revision of *Amblycheila*. Other papers dealing with this genus include Cazier (1939b), Dunn (1980b), Gaumer (1969b), Howden (1970), Knaus (1901), LeConte (1854), Martin (1932), Mateu (1975), Potts (1943), Rivers, (1893), and Snow (1878).

Adult *Omus* have been reported as predators on millipedes (LaBonte and Johnson 1988). Additional papers dealing with *Omus* include Berghe (1990, 1994), Blaisdell and Reynolds (1917), Casey (1897, 1914a, 1924), Cazier (1937a, 1942), G. Horn (1868), W. Horn (1930), Leffler (1979b, 1985a, 1985b [larvae]), Leffler, Nelson, and van den Berghe (1986), Maser (1977b,c), Nunenmacher (1940), Pratt (1939), Reiche (1838), and Ward (1971, 1980).

Tribe Cicindelini

Cicindela Linnaeus, 1758, +99 species, generally distributed.
   subgenus *Cylindera* Westwood, 1831
   subgenus *Dromochorus* Guérin-Méneville, 1845
   subgenus *Ellipsoptera* Dokhtouroff, 1883
   subgenus *Habroscelimorpha* Dokhtouroff, 1883
   subgenus *Cicindela* Linnaeus
   subgenus *Tribonia* Rivalier, 1954
   subgenus *Cicindelidia* Rivalier, 1954
   subgenus *Eunota* Rivalier, 1954
   subgenus *Microthylax* Rivalier, 1954
   subgenus *Opalidia* Rivalier, 1954
   subgenus *Brasiella* Rivalier, 1954
   subgenus *Plectographa* Rivalier, 1954

Many papers have been published on this genus. Rivalier (1954) divided *Cicindela* (*sensu latu*) into twelve genera (here listed as subgenera). A complete bibliography for this genus would occupy many more pages than are in this book. I include here references for identification of adults and larvae, regional keys, catalogs, and articles describing behavior of North American faunas.

6

Species List

Eastern tiger beetles included in this book

There is always a question regarding the best way to list species. Some argue for alphabetical listings, which are easier for the nonexpert. Others argue for the phylogenetic listing, with related species grouped together. I have chosen the latter, using the catalog sequence according to Freitag (1999).

Genus *Megacephala* Latreille 1802
    Megacephala carolina carolina (Linnaeus)
    Megacephala carolina floridana Leng and Mutchler
    Megacephala carolina chevrolati Chaudoir
    Megacephala virginica (Linnaeus)
Genus *Cicindela* Linnaeus 1758
  subgenus *Cicindela* Linnaeus
    Cicindela duodecimguttata Dejean
    Cicindela formosa generosa Dejean
    Cicindela hirticollis hirticollis Say
    Cicindela hirticollis rhodensis Calder
    Cicindela limbalis Klug
    Cicindela longilabris longilabris Say
    Cicindela nigrior Schaupp
    Cicindela patruela patruela Dejean
    Cicindela purpurea purpurea Olivier
    Cicindela repanda repanda Dejean
    Cicindela scutellaris unicolor Dejean
    Cicindela scutellaris lecontei Haldeman
    Cicindela sexguttata Fabricius
    Cicindela splendida Hentz
  subgenus *Tribonia* Rivalier
    Cicindela ancocisconensis T. W. Harris
    Cicindela tranquebarica tranquebarica Herbst

subgenus *Cicindelidia* Rivalier
    Cicindela abdominalis Fabricius
    Cicindela highlandensis Choate
    Cicindela marginipennis Dejean
    Cicindela punctulata Olivier
    Cicindela rufiventris hentzii Dejean
    Cicindela rufiventris rufiventris Dejean
    Cicindela scabrosa Schaupp
    Cicindela trifasciata ascendens LeConte
subgenus *Habroscelimorpha* Dokhtouroff
    Cicindela dorsalis media LeConte
    Cicindela dorsalis saulyci Guérin-Méneville
    Cicindela severa LaFerté-Sénectère
    Cicindela striga LeConte
subgenus *Eunota* Rivalier
    Cicindela togata togata LaFerté-Sénectère
subgenus *Microthylax* Rivalier
    Cicindela olivacea Chaudoir
subgenus *Brasiella* Rivalier
    Cicindela viridicollis Dejean
subgenus *Cylindera* Westwood
    Cicindela cursitans LeConte
    Cicindela unipunctata Fabricius
subgenus *Dromochorus* Guérin-Méneville
    Cicindela pilatei Guérin-Méneville
subgenus *Ellipsoptera* Dokhtouroff
    Cicindela blanda Dejean
    Cicindela cuprascens LeConte
    Cicindela gratiosa Guérin-Méneville
    Cicindela hamata lacerata Chaudoir
    Cicindela hirtilabris LeConte
    Cicindela lepida Dejean
    Cicindela macra macra LeConte
    Cicindela marginata Fabricius
    Cicindela puritana G. Horn
    Cicindela wapleri LeConte

7

Morphological Characters Used for Species Determination

The next few pages illustrate external characters used in the identification of tiger beetle genera *Megacephala* and *Cicindela*. The locations of characters used for species determination are briefly discussed here, with reference to selected plates illustrating the more important ones. Unfamiliar terms can be found in the glossary.

Head (plates 34–41)—Mouthparts, including mandibles (presence or absence of ventral tooth) (plate 37), labrum (number of setae and number and shape of anterior margin teeth), clypeus with or without decumbent seta (plates 35–36). Gena ("cheek") with or without decumbent setae (plates 37, 40); first antennal segment with one or more erect setae; frons with or without setae; vertex of head with or without setae; number of supraorbital setae; pronotum shape.

Elytra (plates 42–50)—Surface smooth or punctate, granulate; color; maculation extent, location, and shape; apex with or without microserrulations (small teeth) (plate 44); apical sutural spine location and shape (plates 42–43); suture with or without adjacent foveae (pits); scutellum present or absent.

Venter (plate 51)—Underside; front and middle trochanters (second leg segment) with or without erect setae (single, tall hairs); sternal sclerites (underside segments) with or without decumbent (pressed against body) setae; legs with covering of decumbent setae (plate 52); lateral sclerites with or without decumbent setae; femur (third leg segment) length; tarsal claw size.

These characters used in conjunction with species photographs will make it possible to identify the eastern species of tiger beetles discussed here.

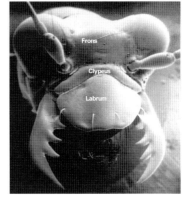

Plate 34. Head of *Cicindela abdominalis* adult.

 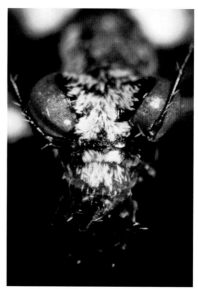

Plate 35. (*above left*) *Cicindela gratiosa* clypeus with many decumbent setae, labrum glabrous except for marginal setae.
Plate 36. (*above right*) *Cicindela hirtilabris* clypeus and labrum covered with many decumbent setae.
Plate 37. (*below left*) *Cicindela marginata* male, mandibular tooth.
Plate 38. (*below right*) *Cicindela sexgutatta* head. Labrum with medial teeth.

Plate 39. *Cicindela scutellaris unicolor* mouthparts.

Plate 40. *Cicindela formosa generosa* gena ("cheek") and lateral view of head.

Plate 41. *Cicindela formosa generosa* dorsal view of head.

Plate 42. Apex of female *Cicindela hamata lacerata* elytron, showing recessed sutural spine.

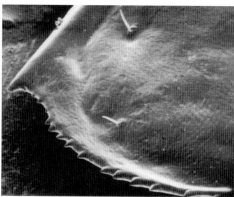

Plate 43. (*left*) *Cicindela marginata* female, deflexed apex of elytron.

Plate 44. (*above*) *Cicindela scabrosa* apical elytral microserrulations.

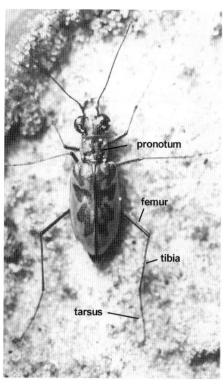
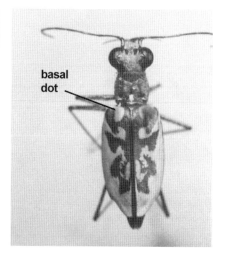

Plate 45. (*left*) *Cicindela blanda* identifying structures used in key.

Plate 46. (*above*) *Cicindela blanda* showing marginal line and apical lunule joined.

Morphological Characters Used for Species Determination · 51

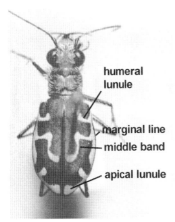

Plate 47. (*far left*) *Cicindela hirticollis* female illustrating components of elytral markings of tiger beetles.

Plate 48. (*left*) *Cicindela repanda* showing typical complete markings.

Plate 49. (*left*) *Cicindela scutellaris unicolor* showing smooth, impunctate elytra (South Carolina).
Plate 50. (*below left*) *Cicindela sexguttata* showing granular elytra (South Carolina).

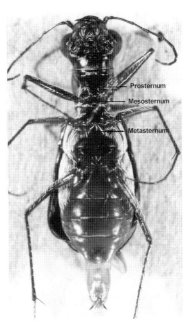

Plate 51. Ventral view of tiger beetle (*C. scabrosa*) showing locations of pro-, meso-, and metasterna.

Plate 52. *Cicindela pilatei* showing decumbent setae on legs.

8

Identification Keys to Eastern Tiger Beetles

If you have never used a key before, the process is a series of choices between two alternative descriptions. I have designed this key to be as straightforward as possible. Each numbered paragraph is one of a pair of choices; each choice leads to another pair of choices, and so on, until a species name is reached. I start with a key to the genera—is it *Megacephala* or *Cicindela*?—each of these then has its own key to species. You can check your outcome against the pictures. If you appear to be incorrect, backtrack by referring to the numbers in parentheses, which refer to the couplet from which you came, to look for a mistake in your choices. Keep working back and forth until you are satisfied you have arrived at the best determination. If you just cannot fit your specimen to the descriptions in the key, it is possible you have found a new species or a species not anticipated for this region. It is also possible that there is a mistake in the key. In any of these events, please contact the author at *pmc@mail.ifas.ufl.edu* or mail to Paul Choate, Bldg. 970, Natural Area Drive, University of Florida, Gainesville, FL 32611.

Key to genera of eastern tiger beetles

1. Scutellum hidden (plate 55); anterior pronotal angles prominent; third segment of maxillary palpus longer than fourth .... *Megacephala*
—Scutellum visible from above (plate 61); anterior pronotal angle not projecting forward; third segment of maxillary palpus shorter than fourth .... *Cicindela*

Key to species of *Megacephala*

1. Dorsal surface dark greenish-black throughout, elytra lacking pale apical lunules (plates 59, 60) .... *virginica* (Linnaeus)
—Elytra with pale apical lunules 2

2(1). Anterior lateral regions of elytra usually black or dark green; violet or coppery reflection may be visible near scutellum; apical lunule with anterior portion divergent (plate 53) .... *carolina floridana* .... (=? *chevrolati* Chaudoir) (plate 57)
—Anterior lateral regions of elytra frequently with extensive violet or coppery reflection; apical lunule with anterior portions convergent (plates 54, 55) .... *carolina carolina* (Linnaeus)

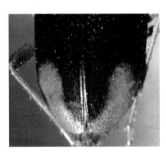 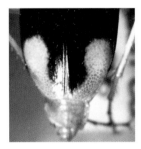

Plate 53. Apical lunule of *Megacephala carolina floridana*.

Plate 54. Apical lunule of *Megacephala carolina carolina*.

Key to eastern species of *Cicindela*

1. Front trochanters with 1 (rarely 2) subapical seta, middle trochanters with or without such seta 8
—Front trochanters lacking subapical setae, middle trochanters also without such setae 2

2(1). Hind femora long, extending more than one-third their length beyond end of body; tarsal claws nearly as long as last tarsal segment; color of abdominal segments variable, rufous coloration restricted to last 1 or 2 segments; white species occurring on coastal beaches (plates 126, 132) 3
—Hind femora short, not extending more than one-third beyond body; tarsal claws much shorter than last tarsal segment; all abdominal segments rufous 4

3(2). Mostly white, largely unmarked elytra; beaches of Gulf coast of Florida from Keys to Alabama (plates 130–132) ....
.... *dorsalis saulcyi* Guérin
—Elytra marked with dark lines (plates 126–129); beaches of east coast south to the Florida Keys .... *dorsalis media* Say

4(2). Elytral tips not microserrulate; markings of elytra connected along lateral margins; abdomen rufous; found on large cobblestone bars in rivers and streams (plates 107–109) .... *marginipennis* Dejean
—Elytral tips microserrulate (plate 44); abdomen color variable 5

5(4). Labrum with 2 anterior medial setae, 2 lateral setae (plate 34) .. 6
—Labrum with 4 anterior medial setae, 2 lateral setae 7

6(5). Ventrally glabrous; lacking lateral pronotal hairs; frequently with greenish reflections; Highlands and Polk counties (plates 103, 104) .... *highlandensis* Choate
—Ventrally lateral sclerites covered with white decumbent setae, also sternites 1–4 with lateral decumbent setae (plate 51); pronotum with at least a few lateral setae, or if absent, punctures visible along suture (plates 100, 101) .... *abdominalis* Fabricius

7(5). Elytra deeply punctured, scabrous; surface shining; only in peninsular Florida, in scrub habitat (plates 120–122) .... *scabrosa* Schaupp
—Elytra shallowly punctate or impunctate; surface dull; known only from Liberty and Gadsden counties, Florida (plates 116–119) .... *rufiventris* Dejean

8(1). Clypeus densely to sparsely clothed with decumbent setae (plate 35) 9
—Clypeus glabrous or with a few erect setae (plate 38) 19

9(8). Prosternum with dense decumbent setae (plate 51) 10
—Prosternum glabrous 14

10(9). Elytra impunctate, dull, mostly white dorsally with sutural area coppery (plate 159) 11
—Elytra deeply punctate, dull to shiny; color variable 12

11(10). Labrum with few (<10) or no decumbent setae (ignore marginal row) (plate 35) .... *gratiosa* Guérin
—Labrum with many (>20) decumbent setae; restricted to peninsular Florida (plate 36) .... *hirtilabris* LeConte

12(10). Sides of pronotum very convex; markings consist of a broad marginal band or elytra almost entirely white (plates 139, 140) .... *togata* Laferté

—Sides of pronotum straight or slightly curved; markings consist of "normal" maculations; coastal marshes and beaches, Gulf and Atlantic coasts 13

13(12). Elytra of female deflexed at tips, apical spine slightly retracted (plate 43); right mandible of male with prominent tooth below (plate 37); east coast and north along Gulf coast extending just north of Horseshoe Beach .... *marginata* Fabricius

—Elytra of female not deflexed at tips, apical spine much retracted (plate 42); right mandible of male with bump or no tooth below; Gulf coast region, from Florida Keys .... *hamata lacerata* Chaudoir

14(9). Appendages depigmented; overall pale species; small size (plate 166) .... *lepida* Dejean

—Appendages pigmented, size and color variable 15

15(14). Middle band of elytral macula long, "normal" (plate 48) 16

—Middle band of elytral macula very short, broad basally, narrow apically (plate 174); small beetles (<11 mm); western panhandle from Shoal River to Alabama (plate 175) .... *wapleri* LeConte

16(15). Middle band of elytral macula slightly sinuate (plates 154, 167, 171) 17

—Middle band of elytral macula very sinuate (plate 151); larger beetle; panhandle from Apalachicola River, west to Alabama ....
.... *blanda* Dejean

17(16). Northeastern states (Chesapeake Bay to New Hampshire); isolated populations on Connecticut River, also Calvert Cliffs, Maryland (plate 171) .... *puritana* Horn

—West and south of Appalachian mountains 18

18(17). Elytra dull, shallowly punctate (plate 167) ...... *macra* LeConte

—Elytra shiny, deeply punctate (plate 154) .... *cuprascens* LeConte

19(8). Frons with erect setae (besides supraorbital setae) (plate 41) . 34

—Frons glabrous or with a few decumbent setae (besides supraorbital setae; there may be a cluster of 10 or more setae near front of eyes) (plates 38, 39) 20

20(19). Legs and tarsi clothed throughout with fine decumbent setae (plate 52); elytra with row of green foveae near suture (plates 148–150) .... *pilatei* Guérin-Menéville
—Legs and tarsi setose, but not clothed with fine decumbent setae .. 21

21(20). Small size (<9 mm long); elytral color brown; prothorax cylindrical with straight sides (plates 212, 213) 22
—Size larger (>12 mm long); other characters variable 23

22(21). Head and pronotum same color as elytra (plate 145) ....
.... *cursitans* LeConte
—Head and pronotum brilliant green, contrasting with brown elytra (plates 143, 144) .... *viridicollis* Dejean

23(21). Proepisternum with setae (may be just a few near the edges of the coxa) 24
—Proepisternum glabrous (plates 146, 147) ...... *unipunctata* Fabricius

24(23). Labrum longer than wide; elytra granulate, dull (plates 73, 74) .... *longilabris* Say
—Labrum medium to short (plates 38, 39) 25

25(24). Elytral tips microserrulate (plate 44) 27
—Elytral tips not microserrulate 26

26(25). Cluster of setae near front of eyes (plate 39); elytra impunctate except for erect seta; without apical pale markings in Florida ....
.... *scutellaris unicolor* Dejean
—With just 2 supraorbital setae near front of eyes (plate 38); pronotum glabrous; elytra shiny, with large deep punctures and scattered pale markings; also apical white lunule (plates 136–138) ... *striga* LeConte

27(25). First antennal segment with 1 sensory seta 29
—First antennal segment with 3–4 sensory setae (plates 78, 90, 91) ....
.... 28

28(27). Elytra granulate; lateral margins of abdomen with sparse decumbent setae; middle band of elytra usually complete (plate 78) ....
.... *patruela* Dejean

—Elytra shallowly to deeply punctate; abdomen glabrous laterally; middle elytral band usually broken into dots or absent (plates 90, 91) .. .. *sexguttata* Fabricius

29(27). Elytral markings complete, often fused 33
—Elytral markings broken into dots or absent; middle band incomplete 30

30(29). Labrum with 8 or more setae (plates 124, 125) .... *trifasciata ascendens* LeConte
—Labrum with <8 setae 31

31(30). Labrum unidentate (1 median tooth) .... *punctulata* Olivier
—Labrum tridentate (3 median teeth) 32

32(31). Elytra with complete humeral lunule or at least a dot on humeral angle (plate 141) .... *olivacea* Chaudoir
—Elytra without dot on humeral angle; markings reduced to apical lunule and broken dots in mid elytra (plate 134) .... *severa* LaFerté-Sénecterè

33(29). Last visible sternite rufous colored; dorsum olive-green and greasy appearing (plate 141) .... *olivacea* Chaudoir
—Abdomen dark brown with metallic reflection; middle band of elytra very sinuate (plate 124) .... *trifasciata ascendens* LeConte

34(19). Genae with setae (plates 37, 40) 35
—Genae glabrous 43

35(34). Labrum with one or no median tooth 36
—Labrum with 3 or more median teeth 37

36(35). Pronotum narrow, front angle rounded; humeral lunule usually complete and connected to or slightly removed from marginal line (plates 80, 81) .... *repanda* Dejean
—Pronotum wide; frontal angles acute; humeral lunule usually broken and widely separated from marginal line (plates 61, 62) .... *duodecimguttata* Dejean

37(35). Humeral lunule complete (plates 47, 48), projecting far mesad

(plate 98), or obliterated by heavy markings (plates 63, 65) 38
—Humeral lunule absent, broken into dots 39

38(37). Markings not connected along margin of elytra; humeral lunule long and oblique (plate 98) .... *tranquebarica* Herbst
—Markings connected along margin of elytra; humeral lunule not long and oblique; larger than normal-sized (20 mm–25 mm) species (plate 65) .... *formosa generosa* Dejean

39(37). First antennal segment glabrous (except for sensory setae) (plates 63, 65) .... *formosa generosa* Dejean
—First antennal segment with few to many setae 40

40(39). Elytra dark greenish brown to reddish brown; all elytral lunules complete (plates 96, 97) .... *ancocisconensis* T. W. Harris
—Elytra purplish to red or green or black; lunules complete, broken, or absent 41

41(40). Head and pronotum much different color than red elytra (plates 92–94) .... *splendida* Hentz
—Head, pronotum, and elytra not differing in overall color 42

42(41). Humeral and apical lunules usually present, humeral lunule may be reduced to a dot (plates 71, 72) .... *limbalis* Klug
—Humeral and apical lunules absent; markings reduced to median markings (plate 79) .... *purpurea* Olivier

43(34). Elytra impunctate, not granulate (plate 49) 44
—Elytra granulate or punctate (plate 50) 45

44(43). Median tooth of labrum smaller than lateral teeth; diameter of penultimate segment of labial palp about 2 times diameter of terminal segment at distal end; uniformly black species (Florida); green individuals occur with black individuals in other portions of range .... *nigrior* Schaupp
—Median tooth of labrum larger than lateral teeth (plate 39); diameter of penultimate segment of labial palp equal to diameter of terminal segment at distal end; unmarked greenish blue species in Florida .... *scutellaris unicolor* Dejean

45(43). First antennal segment glabrous or with 1–2 setae (plates 47, 67–70) .... *hirticollis* Say
—First antennal segment with several erect setae 46

46(45). Humeral lunule complete or with at least a dot on humeral angle (plate 98) .... *tranquebarica* Herbst
—Humeral lunule absent, no dot on humeral angle (plate 79) ....
.... *purpurea* Olivier

9

The Species of Eastern Tiger Beetles

Distributions and Habitats

Each eastern tiger beetle species is illustrated with one or more photographs, including at least one of a pinned specimen and usually a live field photograph of the species in its habitat. A distribution map is also presented showing records from eastern states, without exact localities. Florida species distributions include their county records. Other species

Map 1. Florida counties.

not yet recorded from Florida are shown by their state records only. State and county records merely reflect reports/collection from within those bodies and do not imply distribution throughout. A brief description of habitat is also presented when known.

Megacephala carolina carolina (L.)

Megacephala carolina carolina probably occurs throughout only the northern half of Florida—along rivers, in open pastures, and in disturbed areas. The beetle is nocturnal, spending the day hiding under debris and in cracks in the earth.

In the south, coastal Florida populations may be represented by the similar-appearing *carolina floridana* or *chevrolati*. Ronald Huber (pers. comm.) has suggested that *carolina floridana* and *chevrolati* are

Plate 55. *Megacephala carolina* (typical form).

Plate 56. *Megacephala carolina* (typical form) foraging along shoreline of river in Florida panhandle.

Map 2. (*left*) Florida distribution of *Megacephala carolina carolina* (records from Monroe and Dade counties probably represent *M. carolina floridana*).
Map 3. (*right*) Eastern distribution of *Megacephala carolina carolina*.

two distinct species, both occurring in Florida, but this remains to be determined. He has observed that *chevrolati* appears consistently larger than *floridana*.

Megacephala carolina floridana Leng and Mutchler

This variety was described from Everglade County, in south Florida, and is probably more widespread than collection records indicate. It has been recorded from the Florida Keys, where it may be found at night on alkali mudflats, to Dixie County in the north, where it occurs along with *severa* and *striga*.

Plate 57. *Megacephala carolina floridana* (Dixie County, Florida).

Map 4. Florida distribution of *Megacephala carolina floridana*.

Plate 58. *Megacephala carolina floridana* (Dixie County, Florida).

Megacephala virginica (L.)

This distinctive species frequently occurs with both subspecies of *carolina*, but appears to be much more common and widespread in Florida. A nocturnal species, *virginica* is also a rapid runner.

Plate 59. *Megacephala virginica*

Plate 60. *Megacephala virginica* adult scavenging at night (Steinhatchee, Dixie County, Florida).

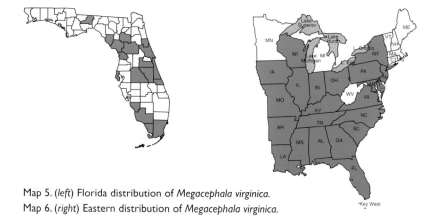

Map 5. (*left*) Florida distribution of *Megacephala virginica*.
Map 6. (*right*) Eastern distribution of *Megacephala virginica*.

Cicindela duodecimguttata Dejean

There are no known records for *duodecimguttata* from Florida. This species may occur with *tranquebarica* or *repanda* in the north.

Map 7. Eastern distribution of *Cicindela duodecimguttata*.

Plate 61. *Cicindela duodecimguttata* (New Hampshire).

Plate 62. *Cicindela duodecimguttata* (New Hampshire).

Plate 63. *C. formosa generosa* (Arkansas).

Cicindela formosa generosa Dejean

There are no known Florida records for *formosa generosa*. This species occurs on large river sandbars and in sandpits.

Map 8. Distribution of *Cicindela formosa generosa*.

Plate 64. New Hampshire. Gravel pit, habitat of *Cicindela formosa generosa*.

Plate 65. *Cicindela formosa generosa* (New Hampshire).

Distributions and Habitats · 67

Plate 66. Photographing *C. formosa generosa* larval burrow.

Cicindela hirticollis Say

This species may be abundant on large river sandbars, but appears to be almost extinct in Florida. A few scattered Florida records are known from the east coast, and on the Escambia River. This is a spring-fall species so may be easily overlooked in Florida.

Plate 67. *Cicindela hirticollis* male (Arkansas).

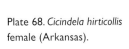

Map 9. Eastern distribution of *Cicindela hirticollis*.

Plate 68. *Cicindela hirticollis* female (Arkansas).

Plate 69. *Cicindela hirticollis rhodensis* (Massachusetts).

Plate 70. *Cicindela hirticollis rhodensis*. Note reduced markings (Massachusetts).

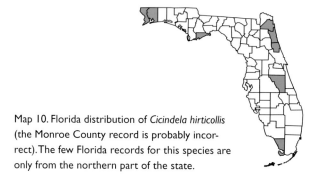

Map 10. Florida distribution of *Cicindela hirticollis* (the Monroe County record is probably incorrect). The few Florida records for this species are only from the northern part of the state.

Cicindela limbalis Klug

Cicindela limbalis is a northern species, found on grass-covered clay banks and along rocky banks of streams, as in Maine.

Plate 71. *Cicindela limbalis*.

Map 11. Eastern distribution of *Cicindela limbalis*.

Plate 72. *Cicindela limbalis*. Not yet recorded from Florida.

Cicindela longilabris Say

Cicindela longilabris is an uncommon northern species that may be found in small patches of open woodland, gravel pits, and boreal forests.

Plate 73. *Cicindela longilabris* (Maine).

Map 12. Eastern Distribution of *Cicindela longilabris*.

Plate 74. *Cicindela longilabris* in "alert" position.

Plate 75. Maine. Habitat of *C. longilabris*.

Cicindela nigrior Schaupp

This species was formerly considered a variety of *C. scutellaris* but Vick and Roman (1985) demonstrated that it should be elevated to species rank. Thus far all Florida records are for the fall of the year. In Georgia (vicinity of Kite), and in South Carolina, green *nigrior* individuals occur with black specimens.

Plate 76. *Cicindela nigrior* (Florida).

Plate 77. *Cicindela nigrior* adult (Kite, Georgia).

Map 13. (*left*) Florida distribution of *Cicindela nigrior*.
Map 14. (*right*) Distribution of *Cicindela nigrior*.

Plate 78. *Cicindela patruela.*

Cicindela patruela Dejean

There are no known Florida records. *C. patruela* is a mountain species in the South.

Map 15. Eastern distribution of *Cicindela patruela.*

Plate 79. *Cicindela purpurea* (Maine).

Cicindela purpurea Olivier

There are no known Florida records for *purpurea*, which is a mountain species in the South. *C. purpurea* is one of the earliest species to emerge in the spring, often missed by collectors because of its tolerance of cold temperatures.

Map 16. Eastern distribution of *Cicindela purpurea.*

Cicindela repanda Dejean

Cicindela repanda is one of the most common species in eastern United States, found on riverbanks, sandbars—apparently always in the vicinity of water. *Repanda* is a diurnal species, apparently never active at night.

Plate 80. *Cicindela repanda*.

Plate 81. *Cicindela repanda* on Escambia River, Florida.

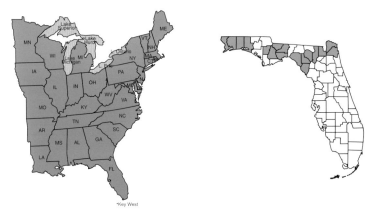

Map 17. (*left*) Eastern distribution of *Cicindela repanda*.
Map 18. (*right*) Florida distribution of *Cicindela repanda*.

Plate 82. *Cicindela scutellaris unicolor* (Florida).

Cicindela scutellaris unicolor Dejean

This species occurs throughout Florida on sand ridges and in scrub areas. Adults occur in spring and fall, and may be active on warm midwinter days. Adults are strictly diurnal. Florida populations are unmarked, but Georgia specimens may show apical markings.

Map 19. Florida distribution of *Cicindela scutellaris unicolor*.

Plate 83. *Cicindela scutellaris unicolor* (Florida).

Plate 84. *Cicindela scutellaris unicolor* (Florida).

Map 20. Distribution of *Cicindela scutellaris unicolor*.

Distributions and Habitats · 75

Variation in eastern *Cicindela scutellaris*

slightly marked at elytral apex, southern variant (plate 85); fully marked, northeastern states (plate 86); totally unmarked form, Florida (plate 87).

Plate 85. *Cicindela scutellaris unicolor* (Wedgefield, South Carolina).

Plate 86. (*left*) *Cicindela scutellaris lecontei* (New Hampshire).
Plate 87. (*right*) *Cicindela scutellaris unicolor* (Florida).

Cicindela sexguttata Fabricius

Cicindela sexguttata is uncommon in Florida. *Sexguttata* is a woodland path species, quick to take flight or hide in grass, where it is well camouflaged.

Plate 88. *Cicindela sexguttata* hiding in grass.

Plate 89. *Cicindela sexguttata*, showing granular elytra.

Distributions and Habitats · 77

Plate 90. (*left*) *Cicindela sexguttata*, immaculate form.
Plate 91. (*right*) *Cicindela sexguttata*, maculate form.

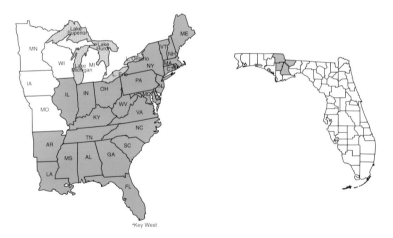

Map 21. (*left*) Eastern distribution of *Cicindela sexguttata*.
Map 22. (*right*) Florida distribution of *Cicindela sexguttata*.

Plate 92. *Cicindela splendida*, illustrating difference in color between elytra and head and pronotum.

Cicindela splendida Hentz

There are no Florida records for this species. *Cicindela splendida* occurs throughout the Appalachian Mountains, as far south as northern Georgia. It is found in habitats similar to those of C. *rufiventris:* roadside clay grassy areas, at higher elevations. A spring-fall species, it occurs as early as March and into September and October.

The head and pronotum of *splendida* are bright green or blue-green, contrasting sharply with the maroon color of the elytra. This distinguishes them from the similar-appearing *purpurea* and *limbalis*, both of which have head and pronotum the same color as the elytra.

Plate 93. *Cicindela splendida* with complete markings (Clayton, Georgia).

Plate 94. Cicindela splendida hiding in pine litter.

Plate 95. Cicindela splendida in "alert" position (Clayton, Rabun County, Georgia).

Map 23. Eastern distribution of Cicindela splendida.

Cicindela ancocisconensis Harris

This species occurs as far south as northern Georgia in the mountains, where it is found on sandbars of rivers and streams.

Plate 96. *Cicindela ancocisconensis* (North Conway, New Hampshire).

Map 24. Distribution of *Cicindela ancocisconensis*.

Plate 97. *Cicindela ancocisconensis* (Ammonoosuc River, New Hampshire).

Cicindela tranquebarica Herbst

Uncommon in Florida, with no known recent collection records. In South Carolina and Georgia I have collected this species in disturbed areas along roadsides that cut through pine forests.

Plate 98. *Cicindela tranquebarica* (Georgia).

Plate 99. *Cicindela tranquebarica* (New Hampshire).

Map 25. (*left*) Eastern distribution of *Cicindela tranquebarica*.
Map 26. (*right*) Florida distribution of *Cicindela tranquebarica*. Based on literature records (Leng, 1915). No specimens seen; no recent collections known.

Cicindela abdominalis Fabricius

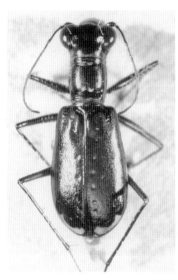

Plate 100. *Cicindela abdominalis*.

Cicindela abdominalis appears in early summer and disappears by early fall. *Abdominalis* overwinters as late instar larvae, pupating in the spring. There is one generation per year. This species is restricted to sand ridges in peninsular Florida, where dominant vegetation is turkey oak (*Quercus laevis*), saw palmetto (*Serenoa repens*), and in some areas Florida rosemary (*Ceratiola ericoides*). Adults and larvae occupy the same habitat, open patches among trees and fire lanes.

Abdominalis seasonally shares the same habitat with *C. hirtilabris* in peninsular Florida and with *C. gratiosa* in the Florida panhandle. In spring and fall and warmer days in winter adults of *C. scutellaris unicolor* also occupy the same habitat. Where *abdominalis* habitat meets roadsides, individuals of *C. punctulata* may also occur during summer months.

Abdominalis is active only during the day and not collected at lights. During the hottest part of the day individuals will seek shade under leaves and debris and at the edges of woods. Not a strong flier, but its small size makes detection difficult. The collector must make frequent backtracks to flush individuals from cover. It is not unusual to think that this species is absent from apparently good habitat, but persistent coverage of area will usually produce specimens.

Larvae occur in the same habitat as adults; the last instar occupies a vertical burrow that may extend 30" deep. The only sure way to exca-

Plate 101. *Cicindela abdominalis* in "alert" position.

Plate 102. Route 26, West of Newberry, Florida. Habitat of *Cicindela abdominalis, hirtilabris,* and *scutellaris unicolor.*

vate these is to insert a long stem of grass into the burrow, then follow the burrow carefully with a shovel. Individuals are frequently found parasitized by larvae of small wasps.

This species was considered to have one subspecies (*scabrosa* Schaupp) and several synonyms. Choate (1984) elevated *scabrosa* Schaupp to species level, described a new species from central Florida (*highlandensis*), and treated *floridana* Cartwright as a synonym of *scabrosa* Schaupp. *Cicindela abdominalis* occurs in the Piedmont of southeastern United States, as far north as New Jersey in the Pine Barrens. Specimens from the northern portion of its range are larger and more extensively maculated than specimens from the southern portion of the range.

Selected Reference

Choate, P. M. 1984. A new species of *Cicindela* Linnaeus (Coleoptera: Cicindelidae) from Florida, and elevation of *C. abdominalis scabrosa* Schaupp to species level. *Entomological News* 95: 73–82.

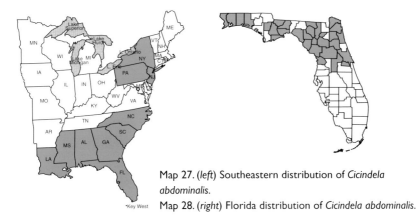

Map 27. (*left*) Southeastern distribution of *Cicindela abdominalis.*

Map 28. (*right*) Florida distribution of *Cicindela abdominalis.*

Cicindela highlandensis Choate

Known only from Polk and Highlands counties in central Florida, this species was newly discovered in 1984. Its restricted range and the decreasing availability of its habitat has made this a candidate for endangered species status.

Selected reference

Choate, P. M. 1984. A new species of *Cicindela* Linnaeus (Coleoptera: Cicindelidae) from Florida, and elevation of C. *abdominalis scabrosa* Schaupp to species. *Entomological News* 95(3):73–82.

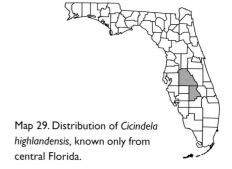

Plate 103. *Cicindela highlandensis* (Highlands County, Florida).

Map 29. Distribution of *Cicindela highlandensis*, known only from central Florida.

Plate 104. *Cicindela highlandensis* (Highlands County, Florida).

Plate 105. Josephine Creek, Highlands County, Florida. Type locality of *Cicindela highlandensis* Choate.

Plate 106. *Cicindela highlandensis* (Highlands County, Florida).

Cicindela marginipennis Dejean

Cicindela marginipennis is known as far south as northern Alabama, where it occurs on rivers with cobblestone bars or coarse gravel.

Plate 107. *Cicindela marginipennis* (Walpole, New Hampshire).

86 · Florida and Eastern U.S. Tiger Beetles

Plate 108. *Cicindela marginipennis* (Alabama).

Plate 109. *Cicindela marginipennis* (New Hampshire).

Map 30. Known distribution of *Cicindela marginipennis*.

Plate 110. Connecticut River, Walpole, New Hampshire. Habitat of *Cicindela marginipennis*.

Cicindela punctulata Olivier

By far the most common summer species, *punctulata* is found along sidewalks, roadsides, and disturbed areas throughout the United States. This species is active at night as well as during the day and comes readily to lights.

Plate 111. *Cicindela punctulata*.

Plate 112. *Cicindela punctulata* (Florida).

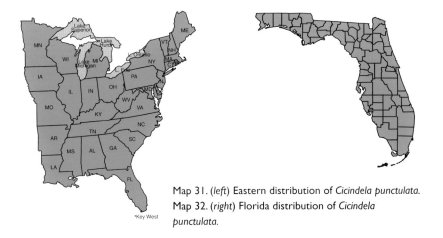

Map 31. (*left*) Eastern distribution of *Cicindela punctulata*.
Map 32. (*right*) Florida distribution of *Cicindela punctulata*.

Cicindela rufiventris hentzii Dejean

This subspecies of *rufiventris* is known only from the hills around Boston, where it occurs on rock outcropping.

Plate 113. *Cicindela rufiventris hentzii* (Massachusetts).

Map 33. Distribution of *Cicindela rufiventris hentzii*.

Plate 114. *Cicindela rufiventris hentzii*. This subspecies is restricted to areas surrounding Boston, especially Blue Hill Reservation, Massachusetts.

Distributions and Habitats · 89

Plate 115. *Cicindela rufiventris hentzii* in "alert" position (Massachusetts).

Cicindela rufiventris rufiventris Dejean

Cicindela rufiventris is known from only two Florida panhandle counties, where it occurs on red-clay banks and along roadsides through pine woods.

Plate 116. (*left*) *Cicindela rufiventris* (Florida).
Plate 117. (*right*) *Cicindela rufiventris*, ventral view showing rufous abdomen.

90 · Florida and Eastern U.S. Tiger Beetles

Plate 118. *Cicindela rufiventris* on woodland path (Georgia).

Plate 119. *Cicindela rufiventris* on eroded clay bank. (Georgia).

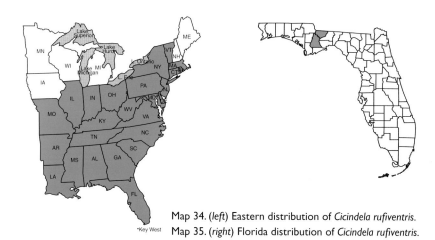

Map 34. (*left*) Eastern distribution of *Cicindela rufiventris*.
Map 35. (*right*) Florida distribution of *Cicindela rufiventris*.

Cicindela scabrosa Schaupp

This species was once considered a variety of *abdominalis*, but was elevated to species rank in 1984. It occurs only in peninsular Florida, in sandpine scrub and along roads that cut through pine flatwoods. *Scabrosa* is strictly diurnal, hiding during the hottest part of daylight hours and difficult to locate because of its small size and secretive nature.

Selected references

Choate, P. M. 1984. A new species of *Cicindela* Linnaeus (Coleoptera: Cicindelidae) from Florida, and elevation of C. *abdominalis scabrosa* Schaupp to species. *Entomological News* 95(3):73–82.

Roman, S. J. 1988. Collecting *Cicindela scabrosa* Schaupp, with notes on its habitat. *Cicindela* 20(2):31–34.

Plate 120. *Cicindela scabrosa.*

Plate 121. *Cicindela scabrosa.*

Plate 122. *Cicindela scabrosa* in "alert" position.

Plate 123. Route 347, north of Lukens, northwest of Cedar Key, Levy County, Florida. Habitat of *Cicindela scabrosa, hirtilabris,* and *scutellaris unicolor.*

Map 36. Distribution of *Cicindela scabrosa*, known only from peninsular Florida.

Cicindela trifasciata ascendens LeConte

A widespread species in the south, *trifasciata ascendens* occurs in a variety of habitats: along lakeshores, riverbanks, and salt mudflats. Found during the winter months on warm days and is active at night as well as daytime, coming to lights.

Plate 124. *Cicindela trifasciata ascendens.*

Plate 125. *Cicindela trifasciata ascendens* (Florida).

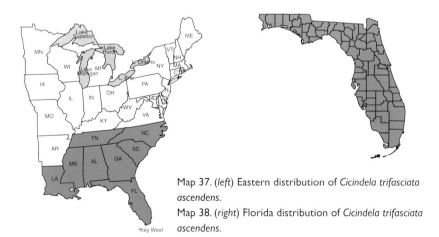

Map 37. (*left*) Eastern distribution of *Cicindela trifasciata ascendens.*

Map 38. (*right*) Florida distribution of *Cicindela trifasciata ascendens.*

Cicindela dorsalis media LeConte

This subspecies occurs along the east coast of the United States from Maryland to south Florida, where it is active at night as well as in daytime. An environmentally sensitive species, populations have been extirpated by vehicular and pedestrian traffic along many miles of shoreline. This subspecies is normally well marked throughout its range.

Plate 126. *Cicindela dorsalis media* (Little Talbot State Park, Duval County, Florida).

Plate 127. *Cicindela dorsalis media* (Florida).

Plate 128. Mating pair *Cicindela dorsalis media* (Little Talbot State Park, Duval County, Florida).

Distributions and Habitats · 95

Plate 129. Little Talbot State Park, Duval County, Florida. *Cicindela dorsalis media* larval burrows are indicated by red flags.

Map 39. Eastern distribution of *Cicindela dorsalis media*.

Map 40. Florida distribution of *Cicindela dorsalis media*.

Cicindela dorsalis saulcyi Guérin-Méneville

This subspecies occurs along the Gulf Coast states, from west coast Florida to Mississippi. Specimens are unmarked (plate 130) in southwest Florida, but are frequently marked (plate 131) in the Florida panhandle. Since these markings are used to distinguish the various subspecies of *dorsalis* this needs to be investigated. This subspecies is defined by its reduced markings and by its Gulf Coast distribution.

Plate 130. (*left*) *Cicindela dorsalis saulcyi*, typical markings (Florida).
Plate 131. (*right*) *Cicindela dorsalis saulcyi* (Florida).

Plate 132. *Cicindela dorsalis saulcyi* (Carabelle Beach, Wakulla County, Florida).

Distributions and Habitats · 97

Plate 133. Alligator Point, Wakulla County, Florida. Habitat of *Cicindela dorsalis saulcyi*.

Map 41. Distribution of *Cicindela dorsalis saulcyi*, restricted to Gulf coast.

Map 42. Florida distribution of *Cicindela dorsalis saulcyi*.

Cicindela severa LaFerté-Sénectère

Cicindela severa occurs in Florida from the Keys to Dixie County on the west coast. Adults are active day and night, come readily to lights, and may be found on alkali mudflats. Where *severa* occurs in Florida one frequently also finds *striga*.

Plate 134. *Cicindela severa* (north of Horseshoe Beach, Dixie County, Florida).

Plate 135. *Cicindela severa* at night (south of Steinhatchee, Dixie County, Florida).

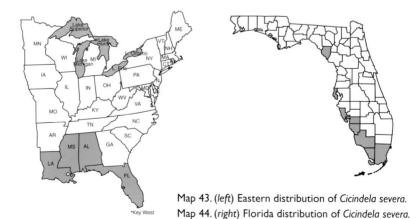

Map 43. (*left*) Eastern distribution of *Cicindela severa*.
Map 44. (*right*) Florida distribution of *Cicindela severa*.

Cicindela striga LeConte

Cicindela striga is considered by tiger beetle collectors to be one of the most desirable Florida species due to difficulty in locating and collecting specimens, yet *striga* remains an anomaly. While adults may be collected in numbers at night at lights, its true habitat remains a mystery. Perhaps this species spends the daylight hours hiding in cracks in soil of coastal mudflats. The large head, reduced eyes, and narrow pronotum are reminiscent of *Megacephala*. Larvae are undescribed. Old literature records mention the appearance of *striga* at campfires; a visit to the type locality at Lake Harney, Florida, revealed habitat quite different from that of coastal locations where it is normally found now.

Plate 136. *Cicindela striga* (Flamingo, Monroe County, Florida).

Plate 137. *Cicindela striga* at night (south of Steinhatchee, Dixie County, Florida).

100 · Florida and Eastern U.S. Tiger Beetles

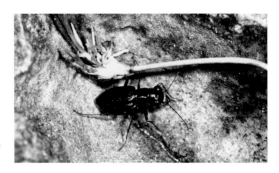

Plate 138. *Cicindela striga* at night (south of Steinhatchee, Dixie County, Florida).

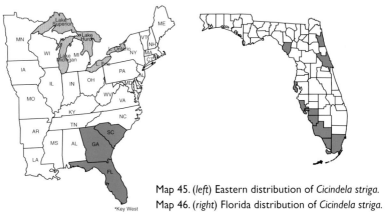

Map 45. (*left*) Eastern distribution of *Cicindela striga*.
Map 46. (*right*) Florida distribution of *Cicindela striga*.

Cicindela togata LaFerté-Sénectère

Cicindela togata is known from two counties in Florida (maps 47, 48). *Togata* occurs on the more sandy parts of coastal mudflats. Adults are active at night and during the daytime.

Plate 139. *Cicindela togata* (Florida).

Plate 140. *Cicindela togata* at night (south of Steinhatchee, Dixie County, Florida).

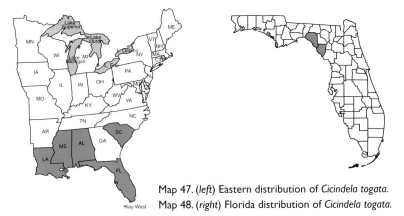

Map 47. (*left*) Eastern distribution of *Cicindela togata*.
Map 48. (*right*) Florida distribution of *Cicindela togata*.

Cicindela olivacea Chaudoir

Cicindela olivacea is a Cuban species, formerly with numerous established populations on the outer Florida Keys (beyond the Seven-Mile bridge). Most known collecting sites have been destroyed by development, and recent collecting trips to Grassy Key, once known to have sizeable populations, have failed to detect this species. Small populations may exist on coral shores of protected locations such as Big Pine Key. All previous known locations have been on the Gulf of Mexico side of the Keys. *Olivacea* is attracted to lights; larvae have not been located in Florida sites.

Plate 141. *Cicindela olivacea* (Grassy Key, Monroe County, Florida).

102 · Florida and Eastern U.S. Tiger Beetles

Plate 142. Grassy Key, Monroe County, Florida. Habitat of *Cicindela olivacea*.

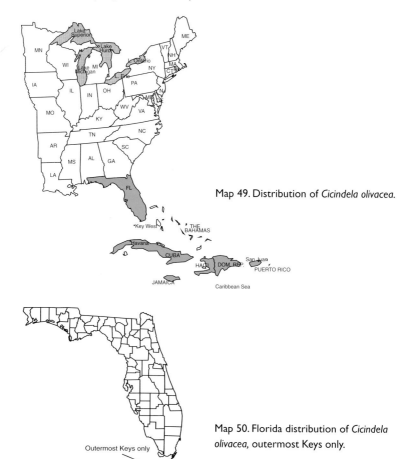

Map 49. Distribution of *Cicindela olivacea*.

Map 50. Florida distribution of *Cicindela olivacea*, outermost Keys only.

Plate 143. (*left*) *Cicindela viridicollis* (Bayamos, Cuba).
Plate 144. (*right*) *Cicindela viridicollis*, ventral view (Bayamos, Cuba).

Cicindela viridicolllis Dejean

The Florida record for this species is based on the capture of a single specimen in the Florida Keys by William Cross, now deceased. Dr. Cross was an avid collector who was murdered while on a collecting trip in Mexico. His collection is believed housed at Mississippi State University.

Reported by Leng and Mutchler (1916) as "running along paths through grassy fields. It flies weakly and, while flying, the brilliant green head and thorax are so conspicuous as to suggest a small bee rather than a *Cicindela*."

Its small size (8 mm in length) and poor flight capability may explain the apparent lack of collections of this species in Florida, or Dr. Cross's specimen may represent a stray from Cuba. Whether or not this species is established in Florida remains to be demonstrated.

Selected Reference

Leng, C. W., and A. J. Mutchler. 1916. Descriptive catalogue of West Indian Cicindelinae. *Bulletin of the American Museum of Natural History* 35: 681–699.

Map 51. Distribution of *Cicindela viridicollis*.

Plate 145. *Cicindela cursitans*.

Cicindela cursitans LeConte

There are no known Florida records for this species. *Cursitans* is said to occur in the vicinity of muddy banks of rivers. Its small size and coloration make it difficult to spot.

Map 52. Eastern distribution of *Cicindela cursitans*.

Cicindela unipunctata Fabricius

A woodland species, *unipunctata* is found along paths and roads and also running about in leaf litter, where it may be collected with pitfall traps. This species is reluctant to fly and may often be hand collected. In the mountains of Georgia it frequently occurs with *rufiventris* on dirt roads that transect clay hillsides.

Plate 146. *Cicindela unipunctata* (Unicoi State Park, White County, Georgia).

Plate 147. *Cicindela unipunctata* in woodland path.

Map 53. Distribution of *Cicindela unipunctata*.

Cicindela pilatei Guérin-Méneville

A flightless species, *pilatei* may be found during the daytime along ditches and creeks that flow through woods. There are no known Florida records.

Plate 148. *Cicindela pilatei* (Lake Charles, Louisiana).

Map 54. Eastern distribution of *Cicindela pilatei,* found also in eastern Texas.

Plate 149. *Cicindela pilatei* (Lake Charles, Louisiana).

Plate 150. Mating pair of *Cicindela pilatei* (Lake Charles, Louisiana).

Cicindela blanda Dejean

A riparian species, *blanda* is found in Florida from the Apalachicola River in Liberty County west to the Escambia River. It occurs on sandbars, is active at night as well as daytime, and comes readily to lights. This species may occur with *wapleri*, especially on the Blackwater River in Florida.

Plate 151. *Cicindela blanda* (Florida).

Plate 152. *Cicindela blanda* (Florida).

Plate 153. Blackwater State Park, Santa Rosa County, Florida. Habitat of *Cicindela blanda*.

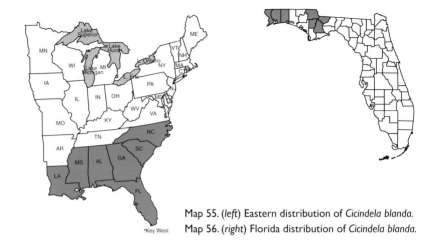

Map 55. (*left*) Eastern distribution of *Cicindela blanda*.
Map 56. (*right*) Florida distribution of *Cicindela blanda*.

Cicindela cuprascens LeConte

There are no known Florida records for *cuprascens*, which occurs on sandbars of larger rivers throughout its range. *Cuprascens* may frequently be found with *macra*, with which it has been confused in many collections; for this reason literature records need to be verified.

Plate 154. *Cicindela cuprascens* (Arkansas).

Map 57. Eastern distribution of *Cicindela cuprascens*.

Cicindela gratiosa Guérin-Méneville

Cicindela gratiosa is a scrub-inhabiting species, found in Florida only in the panhandle, barely extending into the north central part of the state. It occurs with *abdominalis* and *scutellaris unicolor*. *Gratiosa* is replaced by *hirtilabris* in peninsular Florida.

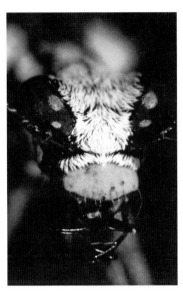

Plate 155. (*left*) *Cicindela gratiosa* (Panacea, Florida).
Plate 156. (*right*) Head of *Cicindela gratiosa,* showing glabrous labrum.
Plate 157. (*below*) *Cicindela gratiosa* (Panacea, Florida).

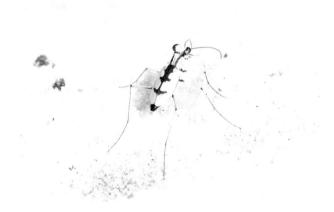

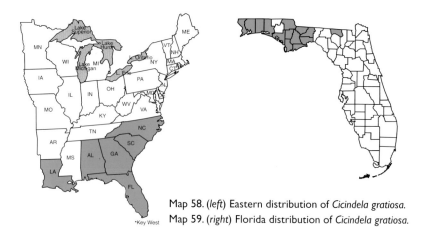

Map 58. (*left*) Eastern distribution of *Cicindela gratiosa*.
Map 59. (*right*) Florida distribution of *Cicindela gratiosa*.

Cicindela hamata lacerata Chaudoir

Hamata lacerata is distributed along the west coast of Florida in tidal salt marshes and on coastal mudflats. It may also occur with *marginata*, *severa*, and *togata*. Adults are active at night as well as during the daytime, and come readily to lights. Specimens are frequently misidentified as *marginata*, and one needs to examine the specimens using key characters—male mandibles and apex of elytra in females (plate 42)—as these species are difficult to identify in the field.

Plate 158. (*left*) *Cicindela hamata lacerata* male (Florida).
Plate 159. (*right*) *Cicindela hamata lacerata* female (Florida).

Distributions and Habitats · 111

Plate 160. Shell Mound, Levy County, Florida. Tidal mudflat exposed at low tide. Habitat of *Cicindela hamata lacerata* adults.

Plate 161. Larval burrow of *Cicindela hamata lacerata* near Cedar Key, Levy County, Florida.

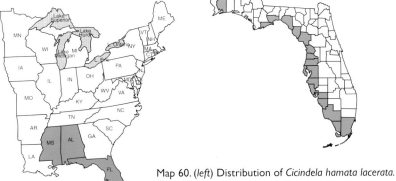

Map 60. (*left*) Distribution of *Cicindela hamata lacerata*.
Map 61. (*right*) Florida distribution of *Cicindela hamata lacerata*.

Cicindela hirtilabris LeConte

This species is restricted to peninsular Florida and extreme southeastern Georgia (St. Simon's Island). It occurs in scrub habitat, frequently with *abdominalis* or *scabrosa*. Adults are active during daytime and at night and will come to lights.

Plate 162. (*left*) *Cicindela hirtilabris* (Alachua County, Florida).
Plate 163. (*right*) *Cicindela hirtilabris* clypeus and labrum with dense covering of decumbent setae.

Plate 164. *Cicindela hirtilabris* (Alachua County, Florida).

Distributions and Habitats · 113

Plate 165. *Cicindela hirtilabris* (Alachua County, Florida).

Map 62. Distribution of *Cicindela hirtilabris*.

Map 63. Florida distribution of *Cicindela hirtilabris*.

Plate 166. *Cicindela lepida*.

Cicindela lepida Dejean

There are no known Florida records for this species. *Lepida* is reported from sandbars and sandpits, where its small size and pale coloration make it difficult to see. Adults are active at night and will come to lights; this may be a more reliable way to collect this species.

Map 64. Distribution of *Cicindela lepida*.

Cicindela macra LeConte

Cicindela macra occurs on sandbars in larger rivers, and adults come to lights. There are no known Florida records. It is often confused with *cuprascens*, and literature records are unreliable for this reason.

Plate 167. *Cicindela macra* (Oklahoma).

Map 65. Eastern distribution of *Cicindela macra*.

Cicindela marginata Fabricius

This species occurs in tidal salt marshes and on coastal mudflats from Maine southward to the West Indies, around the tip of Florida, and as far north as Taylor County on the Gulf coast. Specimens are difficult to separate in the field when they occur with *hamata lacerata* due to their similarity in size and color. But *marginata* can be identified by the elytral tip in females and the mandibular tooth in males (plate 43). Adults are active at night, coming readily to lights in large numbers.

Plate 168. (*left*) *Cicindela marginata* male.
Plate 169. (*right*) *Cicindela marginata* female.

Plate 170. *Cicindela marginata* (Anastasia State Park, St. Johns County, Florida).

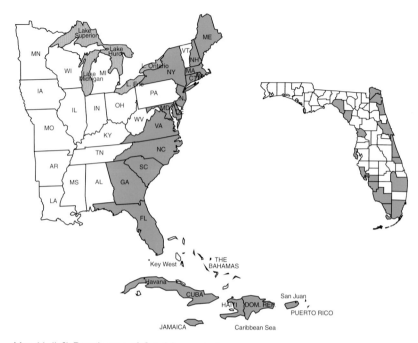

Map 66. (*left*) Distribution of *Cicindela marginata*.
Map 67. (*right*) Florida distribution of *Cicindela marginata*.

Cicindela puritana Horn

This species is considered endangered throughout most of its range. It frequents large eroded banks of rivers and bays, but has disappeared with the building of dams in many of the rivers it once frequented. Remnant populations have recently been discovered on the Connecticut River, where historically it was once widespread.

Plate 171. *Cicindela puritana* (Maryland).

Distributions and Habitats · 117

Plate 172. Calvert Cliffs, Maryland. Habitat of *Cicindela puritana*.

Plate 173. *Cicindela puritana* (Calvert Cliffs, Maryland).

Map 68. Distribution of *Cicindela puritana*.

Cicindela wapleri LeConte

This species seems to prefer muddier riverbanks than *blanda*, with which it may occur in the western panhandle of Florida. Adults are active day and night and may be collected with lights.

Plate 174. *Cicindela wapleri* (Blackwater River State Park, Santa Rosa County, Florida).

Plate 175. *Cicindela wapleri* (Blackwater River State Park, Santa Rosa County, Florida).

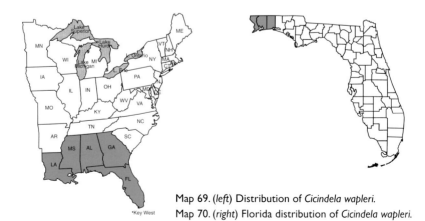

Map 69. (*left*) Distribution of *Cicindela wapleri*.
Map 70. (*right*) Florida distribution of *Cicindela wapleri*.

10

Photographic Catalog of the Eastern Species of *Megacephala*

Plate 176. (*above left*) *Megacephala carolina carolina*.
Plate 177. (*above right*) *Megacephala carolina floridana*.
Plate 178. (*left*) *Megacephala virginica*.

11

Photographic Catalog of the Eastern Species of *Cicindela*

Plate 179. (*left*) *Cicindela duodecimguttata.*
Plate 180. (*right*) *Cicindela formosa.*

Plate 181. (*above left*) *Cicindela hirticollis* female.
Plate 182. (*above right*) *Cicindela hirticollis* male.
Plate 183. (*below left*) *Cicindela purpurea*.
Plate 184. (*below right*) *Cicindela splendida* (Georgia).

Photographic Catalog of the Eastern Species of *Cicindela* · 123

Plate 185. (*above left*) *Cicindela limbalis*. Note that in both *C. limbalis* and *C. purpurea* the head and pronotum are the same color as the elytra, while in *C. splendida* the head and pronotum are a different color than the elytra.
Plate 186. (*above right*) *Cicindela patruela*.
Plate 187. (*below left*) *Cicindela longilabris*.
Plate 188. (*below right*) *Cicindela sexguttata* (Florida).

Plate 189. (*above left*) *Cicindela sexguttata* (Georgia).
Plate 190. (*above right*) *Cicindela sexguttata*, maculate specimen.
Plate 191. (*below left*) *Cicindela repanda*.
Plate 192. (*below right*) *Cicindela nigrior*.

Photographic Catalog of the Eastern Species of *Cicindela* · 125

Plate 193. (*above left*) *Cicindela scutellaris unicolor* (Wedgefield, South Carolina).
Plate 194. (*above right*) *Cicindela scutellaris unicolor* (Florida).
Plate 195. (*below left*) *Cicindela ancocisconensis*.
Plate 196. (*below right*) *Cicindela tranquebarica*.

Plate 197. (*above left*) *Cicindela abdominalis*.
Plate 198. (*above right*) *Cicindela highlandensis*.
Plate 199. (*below left*) *Cicindela scabrosa*.
Plate 200. (*below right*) *Cicindela marginipennis*.

Plate 201. (*above left*) Cicindela punctulata.
Plate 202. (*above right*) Cicindela trifasciata ascendens.
Plate 203. (*below left*) Cicindela rufiventris hentzii.
Plate 204. (*below right*) Cicindela rufiventris (Florida).

128 · Florida and Eastern U.S. Tiger Beetles

Plate 205. (*above left*) *Cicindela dorsalis saulcyi*.
Plate 206. (*above right*) *Cicindela dorsalis media*.
Plate 207. (*below left*) *Cicindela dorsalis saulcyi* (Carabelle Beach, Wakulla County, Florida).
Plate 208. (*below right*) *Cicindela togata*.

Plate 209. (*above left*) *Cicindela severa*.
Plate 210. (*above right*) *Cicindela striga*.
Plate 211. (*below left*) *Cicindela olivacea* (Grassy Key, Monroe County, Florida).
Plate 212. (*below right*) *Cicindela viridicollis* (Cuba).

Plate 213. (*above left*) *Cicindela cursitans*.
Plate 214. (*above right*) *Cicindela unipunctata*.
Plate 215. (*below left*) *Cicindela pilatei* (Lake Charles, Louisiana).
Plate 216. (*below right*) *Cicindela blanda*.

Photographic Catalog of the Eastern Species of *Cicindela* · 131

Plate 217. (*above left*) *Cicindela cuprascens* (Arkansas).
Plate 218. (*above right*) *Cicindela macra* (Oklahoma).
Plate 219. (*below left*) *Cicindela wapleri.*
Plate 220. (*below right*) *Cicindela puritana.*

Plate 221. (*above left*) *Cicindela lepida*.
Plate 222. (*above right*) *Cicindela gratiosa*.
Plate 223. (*below left*) *Cicindela hirtilabris*.
Plate 224. (*below right*) *Cicindela marginata* male.

Plate 225. (*above left*) *Cicindela marginata* female.
Plate 226. (*above right*) *Cicindela hamata lacerata* male.
Plate 227. (*below*) *Cicindela hamata lacerata* female.

Glossary

abdomen: The last, third body region of an insect.
allochrony: Separated in time, populations that occur at different periods are allochronic.
allopatry: Populations separated geographically, not connected, without overlap in range.
antennal insertion: The points where antennae are attached to head.
antennomere: Segment of antenna.
anterior: Toward the front.
apex: At the end.
apical: Toward the end.
apical lunule: White markings at the tip or end of tiger beetle elytra (plate 47).
basal dot: White dot at anterior of wing covers of tiger beetles (plate 46).
Carabidae: Family name of ground beetles.
caudal: Toward the rear.
Cicindela: Genus name of some tiger beetles.
clypeus: Structure immediately above the labrum on the front of an insect's head (plate 34).
Coleoptera: Order name of beetles.
connate: Fused at base or sometimes along entire length.
coxa: First segment of insect leg, attaches leg to body.
decumbent seta: Hair or bristle that is appressed to insect body, not erect.
dichotomous: Two choices, branching; dichotomous keys offer two choices, from which the user is to pick the best choice and go to the next couplet of choices.
disjuction: Distribution of populations that are separate from each other by some barrier.
distal: Toward the outer end.
diurnal: Active in the daytime.

dorsal: Top.
edentate: Lacking teeth.
elytron, *pl.* **elytra:** The hardened front wings of beetles.
emarginate: With a border or edge indented or cut out.
endemic: Peculiar to a given region; native, not introduced; applied to species or higher groups.
episternum: The anterior lateral sclerite between the sternum and notum.
family: In classification hierarchy, the level below an order and above a genus; family names end in *-dae*.
femur, *pl.* **femora:** The third segment of an insect leg; usually the thickest or heaviest part of leg (plate 45).
filiform: Threadlike; pertaining to antennae, segments much longer than wide.
fovea, *pl.* **foveae:** Pit.
frons: The front area of the head between insect eyes (plate 34).
gena: The "cheek" or side of face of an insect, below and behind the compound eyes (plate 40).
genus, *pl.* **genera:** Taxonomic classification below the level of family; first name in a binomial or trinomial scientific name, capitalized and italicized. *Cicindela* is the genus for which the tribe in the family Carabidae was named.
glabrous: Without hair.
granulate: Covered with minute granule elevations, giving the appearance of small bumps.
ground beetle: The common name for beetles in the family Carabidae.
holotype: The single specimen upon which a species name is based.
humeral lunule: Pale crescent-shaped marking on the "shoulder" (anterior outer margin of wings) of tiger beetles (plate 47).
humerus: "Shoulder" of insect as seen from above.
immaculate: Without markings.
intergrade: A hybrid, showing characteristics of two or more forms.
labrum: The "upper lip" of insects (plate 34).
lateral: On or toward the side.
littoral: In intertidal areas along seashores
lunule: Crescent-shaped pale marking on some tiger beetle elytra (plate 47).
maculation: Any pale marking on tiger beetle elytra.
mandible: The jaw containing insect "teeth."

marginal line: Along the outer edge of tiger beetle elytra (plate 47).
maxillary palpus: The outermost palps immediately under the mandibles.
medial: Toward the middle.
Megacephala: The currently accepted name of one tiger beetle genus, originally applied to African species. Many workers consider American members belong to subgenus *Tetracha*. Others believe *Tetracha* deserves full generic status.
mesad: Toward the median plane of the insect body.
meso-: Middle.
meta-: Hind.
microserrulations: Small teeth-like structures at the end of tiger beetle elytra (plate 44).
middle band: White maculation on middle of wing cover, or elytra.
middle tooth: In tiger beetles, tooth on labrum or "upper lip."
monotypic: Having only a single named form.
nocturnal: Active at night.
order: Category of taxonomic classification below class and above family; many order names of insects end in -*ptera*, meaning wings.
parameres: Two lateral processes of genitalia in male Coleoptera.
pars basalis: Together with parameres forms the genital tube in male Coleoptera.
penultimate: Next to last.
pitfall trap: Container set in ground to capture surface-dwelling insects.
polytypic: Having more than one named variety within a species range.
precinctive: Native to and occurring only in a particular location or region.
pro-: Front or first.
proepisternum: The side plate of the prosternum.
pronotum: The "neck," dorsal aspect of first thoracic segment (plate 45).
prosternum: The sclerite between the front legs (plate 51).
prothorax: The first thoracic segments, bears the first pair of legs (plate 45).
proximal: Nearer to the point of attachment, as opposed to **distal.**
punctate: Covered with impressed points or punctures.
riparian: Occurring along streams and rivers.
rufous: Reddish or orange in color.

scabrous: Rough, irregular surface.
sclerite: Any piece of the insect body bounded by sutures.
sclerophyll: Plant types that characterize scrub habitats.
scrub: In Florida, habitat with distinctive oak-pine mixture on well-drained, nutrient-poor sandy soils.
scutellum: Triangular structure at anterior suture where forewings (elytra) meet in *Cicindela*.
semixeric: Slightly more moist habitat, able to support greater plant diversity.
sensu latu: In the broad sense of a term.
sensu strictu: In a narrow, restricted sense of a term.
seta/setae: Erect hairs, longer than other hairs that may be present.
sinuate: S-shaped.
speciation: The process of evolving into a distinct species.
species: A group of individuals capable of interbreeding and producing fertile offspring; in taxonomic classification, second name in a binomial or trinomial scientific name, lowercase, italicized.
sternite: A subdivision of a sternum.
sternum, *pl.* **sterna:** The entire ventral division of any segment.
subapical: A short distance removed from apex of structure.
subspecies: A subdivision of a species, a geographic segregate worthy of a name, capable of interbreeding with other subspecies of a given species.
supraorbital seta: Bristle over eye (plate 41).
suture: The line along which the elytra of beetles meet; also, a depressed line or groove in the insect exoskeleton, demarcates parts of an insect body.
sympatry: Overlapping ranges of insect species.
synonym: A scientific name applied to an insect that has been determined to be the same species as another earlier named insect species.
tarsomere: Segment of tarsus.
tarsus, *pl.* **tarsi:** Leg segment immediately beyond the tibia; the "feet" of insects (plate 45).
tergite: A subdivision of a tergum.
tergum: The upper surface of any body segment.
Tetracha: Considered a synonym of *Megacephala* by some workers, or a separate genus by others.
thorax: The second major body region, containing all pairs of legs.

tibia: Fourth segment of an insect leg, generally the longest, thinnest (plate 45).
tiger beetle: Common name for the beetle family Cicindelidae, or the ground beetle tribe Cicindelini by some workers.
topotype: Specimen that has been collected at the type locality of a species or subspecies.
tridentate: With three teeth.
trochanter: The second segment of an insect leg.
type locality: Locality from which the holotype of a species was described.
ventral: The underside.
vertex: Top of head between the eyes.
vestiture: General surface covering of hairs.
xeric: Dry habitat, soil deficient in moisture to support much plant diversity.

Works Cited and Suggested Readings

Acciavatti, R. E. 1973. Coastal *Cicindela* from interior eastern Texas. *Cicindela* 5(2): 25–27.
———. 1980. A review of *Cicindela praetextata* from the southwest United States (Coleoptera: Cicindelidae). *Southwest. Entomol.* 5(4): 231–244.
Acciavatti, R. E., T. J. Allen, and C. Stuart. 1992. The West Virginia tiger beetles (Coleoptera: Cicindelidae). *Cicindela* 24(3/4): 45–78.
Acciavatti, R. E., and J. C. Bowe. 1991. Father Bernard Rotger: his life, ministry and contributions to entomology. *Cicindela* 23(4): 55–64.
Acciavatti, R. E., C. R. Rotger, and W. A. Iselin. 1979. Checklist of New Mexico *Cicindela* with regional distributions. *Cicindela* 11(2): 27–32.
Acorn, J. H. 1985. The ecology of sand dune tiger beetles. *Univ. Alberta Agric. For. Bull.* 8(2): 39–41.
———. 1988. Mimetic tiger beetles and the puzzle of cicindelid coloration (Coleoptera: Cicindelidae). *Coleopt. Bull.* 42(1): 28–33.
———. 1989. Tiger, tiger burning bright. *Nat. Can. (Ott.)* 18(4): 11–12.
———. 1991. Habitat associations, adult life histories, and species interactions among sand dune tiger beetles in the southern Canadian prairies (Coleoptera: Cicindelidae). *Cicindela* 23(2/3): 17–48.
———. 1994. Tiger beetles (Coleoptera: Cicindelidae) of the Lake Athabasca sand dunes—an intriguing northern assemblage. *Cicindela* 26(1): 9–16.
———. 1992. The historical development of geographic color variation among dune *Cicindela* in western Canada (Coleoptera: Cicindelidae), pp. 217–233. In: G. R. Noonan, G. E. Ball, and N. E. Stork, eds., *The biogeography of ground beetles of mountains and islands*. Andover, Hampshire: Intercept. 256 pp.
Adames, A. J. 1971. A revision of the crabhole mosquitoes of the genus *Deinocerites*. *Contrib. Am. Entomol. Inst.* 24, vol. 7, no. 2. 154 pp.
Adams, C. C. 1901. Baseleveling and its faunal significance, with illustrations from southeastern United States. *Am. Nat.* 35: 839–852.
———. 1902. Southeastern United States as a center of geographical distribution of flora and fauna. *Biol. Bull.* 3: 115–131.
Allee, W. C., A. E. Emerson, O. Park, T. Park, and K. P. Schmidt. 1949. *Principles of animal ecology*. Philadelphia: W. B. Saunders. 837 pp.

Allen, R. T. 1980. A review of the subtribe Myadi: description of a new genus and species, phylogenetic relationships, and biogeography. *Coleopt. Bull.* 34 (1): 1–29.

Altaba, C. R. 1991. The importance of ecological and historical factors in the production of benzaldehyde by tiger beetles. *Syst. Zool.* 40(1): 101–105.

Amadon, D. 1944. A preliminary life history study of the Florida jay, *Cyanocitta c. coerulescens*. *Am. Mus. Novit.* 1252: 1–20.

Anderson, J. F., and G. R. Ultsch. 1987. Respiratory gas concentrations in the microhabitats of some Florida arthropods. *Comp. Biochem. Physiol. A Comp. Physiol.* 88a(3): 585–588.

Andrews, F. G. 1975. Notes on a rare alpine tiger beetle, *Cicindela purpurea plutonica*. *Cicindela* 7(2): 39–40.

Anon. 1989. Listing proposals—October/November 1989. *Endangered Species Tech. Bull.* 14(11–12): 4–6.

Anon. 1990. Final listing rules approved for four species. *Endangered Species Tech. Bull.* 15(9): 6.

Antevs, E. 1931. Late glacial chronology of North America. *Smithson. Inst. Annu. Rep.* 1931: 313–324.

———. 1947. Biogeographical principles. *Oreg. State Univ. Biol. Colloq.* 8: 7–9.

Antoine, J. W., and T. E. Pyle. 1970. Crustal studies in the Gulf of Mexico. *Tectonophysics* 10: 447–494.

Antoine, M. 1950. Notes d'entomologie morocaine. LII. A propos des *Cicindela campestris* L. et *moroccana* Fab. *Rev. Fr. Entomol.* 17: 285–290.

———. 1951. Sur le demembrement du genre *Cicindela*. *Rev. Fr. Entomol.* 18(2): 88–91.

Antonelli, A. L. 1987. Predacious ground beetles. Extension Bulletin no. 1447, Cooperative Extension, College of Agriculture, Washington State University, Pullman. 3 pp.

Applin, P. L., and E. R. Applin. 1944. Regional subsurface stratigraphy and structure of Florida and southern Georgia. *Am. Assoc. Pet. Geol. Bull.* 28: 1673–1753.

Ashmead, W. H. 1903. Classification of the fossorial, predaceous and parasitic wasps, or the superfamily Vespoidea. Subfamily II. Methocinae. *Can. Entomol.* 35: 155–158.

Atwood, W. W. 1940. *The physiographic provinces of North America*. Boston: Ginn. 536 pp.

Bailey, R. M. 1948. Status relationships, and characters of the Percid fish, *Poecilichthys sagitta* Jordan and Swain. *Copeia* 2: 77–85.

Baker, G. T., and W. A. Monroe. 1995. Sensory receptors on the adult labial and maxillary palpi and galea of *Cicindela sexguttata* (Coleoptera: Cicindelidae). *J. Morphol.* 226(1): 25–31.

Balazuc, J., and F. Chalumeau. 1978. Contribution a la faune des Antilles francaises. Cicindelides (Coleopt. Caraboidea, Cicindelidae). *Nouv. Rev. Entomol.* 8(1): 17–26.

Balduf, W. V. 1935. *Bionomics of entomophagous Coleoptera.* New York: J. S. Swift.

Ball, G. E. 1959. A taxonomic study of the North American Licinini with notes on the old world species of the genus *Diplocheila* Brulle. *Mem. Am. Entomol. Soc.* 16. 258 pp.

———. 1963. The distribution of species of the subgenus *Cryobius* (Coleoptera, Carabidae, *Pterostichus*) with special reference to the Bering land bridge and Pleistocene refugia, pp. 133–151. *In: Symposium, Pacific Basin Biogeography, 10th Pacific Science Congress.* Honolulu: Bishop Museum Press.

Ball, I. R. 1975. Nature and formulation of biogeographical hypotheses. *Syst. Zool.* 24: 407–430.

Ball, M. M., and C. G. A. Harrison. 1969. Origin of the Gulf and Caribbean and implications regarding ocean ridge extension, migration, and shear. *Trans. Gulf Coast Assoc. Geol. Soc.* 19: 287–294.

Barlet, J. 1992. Observations sur le thorax de larves de cicindeles (insectes, coleopteres). *Bull. Soc. R. Sci. Liege* 61(5): 339–349.

Barrera, A. 1962. La peninsula de Yucatan como provincia biotica. *Rev. Soc. Mex. Hist. Nat.* 23: 71–105.

Basilewsky, P. 1966. Revision des *Megacephala* d'Afrique (Coleoptera, Carabidae, Cicindelinae). *Mus. R. Afr. Cent. Tervuren Belg. Ann. Ser. Octavo Sci. Zool.* 152. 150 pp.

Bates, H. W. 1878a. Descriptions of 25 new species of Cicindelidae. *Cistula Entomologica* 2: 329–336.

———. 1878b. On new genera and species of geodephagous Coleoptera from Central America. *Proc. Zool. Soc. Lond.* (1878): 587–609.

———. 1890. Additions to the Cicindelidae fauna of Mexico, with remarks on some of the previously recorded species. *Trans. Entomol. Soc. Lond.* 3: 493–510.

Bauer, K. L. 1991. Observations on the developmental biology of *Cicindela arenicola* Rumpp (Coleoptera: Cicindelidae). *Great Basin Nat.* 51(3): 226–235.

Bauer, T., V. Brauner, and E. Fischerleitner. 1979. The relevance of brightness to visual activity, predation and activity of visually hunting ground-beetles (Coleoptera, Carabidae). *Oecologia* 30: 63–73.

Beard, J. H. 1969. Pleistocene paleotemperature record based on planktonic Foraminifers, Gulf of Mexico. *Trans. Gulf Coast Assoc. Geol. Soc.* 19: 535–553.

Beatty, D. R., and C. B. Knisley. 1982. A description of the larval stages of *Cicindela rufiventris* Dejean (Coleoptera: Cicindelidae). *Cicindela* 14(1–4): 1–17.

Behre, E. H. 1950. Annotated list of the fauna of the Grand Isle region 1928–1946. *Occasional Papers, Louisiana State University Marine Laboratory*, no. 6.

Belkin, J. N., and C. L. Hogue. 1959. A review of the crabhole mosquitoes of the genus *Deinocerites* (Diptera, Culicidae). *Univ. Calif. Publ. Entomol.* 14(6): 411–458.

Bell, R. T., and J. R. Bell. 1962. The taxonomic position of the Rhysodidae. *Coleopt. Bull.* 16(4): 99–106.

———. 1979. Rhysodini of the world Part II. Revisions of the smaller genera. (Coleoptera: Carabidae or Rhysodidae). *Quaest. Entomol.* 15: 377–446.

Benedict, W. 1934. Another *Cicindela*. *Pan-Pac. Entomol.* 10: 76.

Berghe, E. P. van den. 1990. On the habits and habitat of *Omus* (Coleoptera: Cicindelidae). *Cicindela* 22(4): 61–68.

———. 1994. *Omus cazieri*, a new species from southern Oregon (Coleoptera: Cicindelidae). *Cicindela* 26(3–4): 33–39.

Berry, E. W. 1924. The middle and upper Eocene floras of southeastern North America. *U.S. Geol. Surv. Prof. Paper* 92. 29 pp.

———. 1937. Tertiary floras of eastern North America. *Bot. Rev.* 3(1): 31–46.

Bertholf, J. 1983. Tiger beetles of the genus *Cicindela* in Arizona (Coleoptera: Cicindelidae). *Texas Tech Univ. Mus. Spec. Pub.* 19: 1–44.

Bessey, C. E. 1908. The taxonomic aspect of the species. *Am. Nat.* 42: 218–224.

Blackwelder, R. E. 1939. Fourth supplement 1933–1938 (inclusive) to the Leng catalogue of Coleoptera of America, north of Mexico. Mount Vernon, N.Y.: J. D. Sherman, Jr. 146 pp.

———. 1944. Checklist of the Coleopterous insects of Mexico, Central America, the West Indies, and South America. Part 1. *U.S. Natl. Mus. Bull.* 185: 16–20.

Blackwelder, R. E., and R. M. Blackwelder. 1948. Fifth supplement 1939–1947 (inclusive) to the Leng catalogue of Coleoptera, north of Mexico. Mount Vernon, N.Y.: J. D. Sherman, Jr. 87 pp.

Blair, W. F. 1950. Ecological factors in speciation of *Peromyscus*. *Evolution* 4: 253–275.

———. 1958. Distributional patterns of vertebrates in the southern United States in relation to past and present environments, pp. 433–468. *In:* C. L. Hubbs, ed., *Zoogeography*. American Association for the Advancement of Science Publ. 51.

Blaisdell, F. E. 1912. Hibernation of *Cicindela senilis* (Coleopt.). *Entomol. News* 23: 156–159.

Blaisdell, F. E., and L. R. Reynolds, Sr. 1917. A new *Omus*. *Entomol. News* 28: 49–53.

Blanchard, F. N. 1921. The tiger beetles (Cicindelidae) of Cheboygan and Emmet counties, Michigan. *Pap. Mich. Acad. Sci. Arts Lett.* 1: 396–417.

Blatchley, W. S. 1910. An illustrated descriptive catalogue of the Coleoptera or beetles (exclusive of the Rhynchophora) known to occur in Indiana. *Bull. Indiana Dep. Geol. Nat. Res.* 1: 1–1386.

———. 1923. Notes on the Coleoptera of southern Florida, with descriptions of new species. *Can. Entomol.* 55(1): 13–20.

———. 1932. *In days agone: notes on the fauna and flora of subtropical Florida in the days when most of its area was a primeval wilderness.* Indianapolis: Nature Publishing. 338 pp.

Blum, M. S, T. H. Jones, and G. J. House. 1981. Defensive secretions in tiger beetles: cyanogenetic basis. *Comp. Biochem. Physiol. B Comp. Biochem.* 69(4): 903–904.

Bock, W. J. 1963. Evolution and phylogeny in morphologically uniform groups. *Am. Nat.* 97: 265–285.

———. 1969a. Phylogenetic systematics, cladistics, and evolution. *Evolution* 22: 646–648.

———. 1969b. Non-validity of the "phylogenetic fallacy." *Syst. Zool.* 18 (1): 111–115.

Bocquet, C. 1969. Le probleme des formes apparents a distribution contigue. *Bull. Zool. Soc. Fr.* 94(4): 517–526.

Boer, P. J. den. 1970. On the significance of dispersal power for populations of carabid beetles. *Oecologia* 4: 1–28.

Borror, D. J., and D. M. DeLong. 1971. *An introduction to the study of insects.* 3d ed. New York: Holt, Rinehart, and Winston. 812 pp.

Bousquet, Y., and A. Larochelle. 1993. Catalogue of the Geadephaga (Coleoptera: Trachypachidae, Rhysodidae, Carabidae including Cicindelini) of America north of Mexico. *Mem. Entomol. Soc. Can.* 167: 1–397.

Boving, A. C., and F. C. Craighead. 1930. Illustrated synopsis of Coleoptera larvae. *Entomol. Am.* 38: 1–351.

———. 1931. An illustrated synopsis of the principal larval forms of the order Coleoptera. *Entomol. Am.* 11: 1–351.

Boyd, H. P. 1973. Collecting tiger beetles in the Pine Barrens of New Jersey. *Cicindela* 5(1): 1–12.

———. 1975. The overlapping ranges of *Cicindela dorsalis* and *C. d. media,* with notes on the Calvert Cliffs area, Maryland. *Cicindela* 7(3): 55–59.

———. 1978. The tiger beetles (Coleoptera: Cicindelidae) of New Jersey with special reference to their ecological relationships. *Trans. Am. Entomol. Soc.* 104(2): 191–242.

———. 1982. Annotated checklist of Cicindelidae (Coleoptera), the tiger beetles, of North and Central America and the West Indies. Marlton, N.J.: Plexus. 31 pp.

———. 1985. Pitfall trapping Cicindelidae (Coleoptera) and abundance of *Megacephala virginica* and *Cicindela unipunctata* in the pine barrens of New Jersey. *Entomol. News* 96(3): 105–108.

Boyd, H. P., and R. W. Rust. 1982. Intraspecific and geographic variations in *Cicindela dorsalis* Say (Coleoptera: Cicindelidae). *Coleopt. Bull.* 36(2): 221–239.

Brady, A. R. 1972. Geographic variation and speciation in the *Sosippus floridanus* species group. *Psyche* 79(1–2): 27–48.

Bram, A. L, and C. B. Knisley. 1982. Studies of the bee fly, *Anthrax analis* (Bombyliidae), parasite on tiger beetle larvae (Cicindelidae). *Va. J. Sci.* 33: 99.

Braun, E. L. 1955. The phytogeography of unglaciated eastern United States and its interpretation. *Bot. Rev.* 22(6): 297–375.

Briggs, J. C. 1974. Operation of zoogeographic barriers. *Syst. Zool.* 23(2): 248–256.

Brimley, C. S. 1938. *The insects of North Carolina, being a list of the insects of North Carolina and their close relatives.* Raleigh: North Carolina Department of Agriculture, Division of Entomology. 560 pp.

———. 1942. *Supplement to Insects of North Carolina.* Raleigh: North Carolina Department of Agriculture, Division of Entomology. 39 p.

Britton, W. E. 1920. Checklist of the insects of Connecticut. *Conn. Geol. Nat. Hist. Surv. Bull.* 31. 397 pp.

Broecker, W. S. 1968. Mikanovitch hypothesis supported by precise dating of coral reefs and deep-sea sediments. *Science* 159: 297–300.

Bromley, S. W. 1914. Asilids and their prey. *Psyche* 21(6): 192–197.

Brouerius van Nidek, C. M. C. 1978. Mexican Cicindelinae (Coleoptera). *Entomologische Blaetter fuer Biologie und Systematik der Kaefer* 74(1–2): 39–44.

———. 1980. Description of some new Cicindelinae (Col.). *Entomologische Blaetter fuer Biologie und Systematik der Kaefer* 75(3): 129–137.

Brown, C. A. 1938. The flora of Pleistocene deposits in the western Florida parishes, West Feleciana Parish, and East Baton Rouge Parish, La. *State La. Dep. Conserv. Geol. Bull.* 12: 59–96.

Brown, K. S., P. M. Sheppard, and J. R. Turner. 1974. Quaternary refugia in tropical America: evidence from race formations in *Heliconius* butterflies. *Proc. R. Soc. Lond. Ser. B* 187: 369–378.

Brown, W. J. 1959. Taxonomic problems with closely related species. *Annu. Rev. Entomol.* 4: 77–98.

Brown, W. L., and E. O. Wilson. 1953. The case against the trinomen. *Syst. Zool.* 3: 174–176.

———. 1956. Character displacement. *Syst. Zool.* 5: 49–64.

Brundin, L. 1967. Insects and the problem of austral disjunctive distribution. *Annu. Rev. Entomol.* 12: 149–168.

———. 1972. Phylogenetics and biogeography. *Syst. Zool.* 21: 69–79.

Bruner, L. 1901. The tiger beetles of Nebraska. *Proc. Nebr. Acad. Sci.* 7: 97–99.

Bryson, H. R. 1939. The identification of soil insects by their burrow characteristics. *Trans. Kans. Acad. Sci.* 42: 245–253.

Brzoska, D. W. 1976. The tiger beetles of Ohio. Master's thesis, Ohio State University, Columbus. 91 pp.

Bucher, W. H. 1947. Problems of earth deformation illustrated by the Caribbean sea basin. *N.Y. Acad. Sci. Trans.*, ser. 2, 9(3): 98–116.

Burdick, D. J, and M. S. Wasbauer. 1959. Biology of *Methoca californica* Westwood (Hymenoptera: Tiphiidae). *Wasmann J. Biol.* 17(1): 75–88.

Burt, W. H. 1954. The subspecies category in mammals. *Syst. Zool.* 3: 99–104.

Bush, G. L. 1975. Modes of animal speciation. *Annu. Rev. Ecol. Syst.* 6: 339–364.

Butterlin, J. 1953. Donnees nouvelles sur la geologie de la Republique d'Haiti. *Soc. Geol. Fr. Bull.*, ser. 6, 3: 283–292.

Cain, A. J. 1953. Geography, ecology, and coexistence in relation to the biological definition of species. *Evolution* 7: 76–83.

Cain, A. J., and G. A. Harrison. 1960. Phyletic weighting. *Proc. Zool. Soc. Lond.* 135: 1–31.

Calder, E. E. 1922a. Changes of names in *Cicindela*. *Can. Entomol.* 54: 191.

———. 1922b. New Cicindelas of the *fulgida* group (Coleopt.). *Can. Entomol.* 54: 62.

Camin, J. H., and R. R. Sokal. 1965. A method for deducing branching sequences in phylogeny. *Evolution* 19: 311–326.

Camp, W. H. 1947. Distribution patterns in modern plants and the problems of ancient dispersals. *Ecol. Monogr.* 17: 159–183.

Camp, W. H., and C. L. Gilly. 1943. The origin and structure of species. *Brittonia* 4: 323–385.

Carlquist, S. 1966. The biota of long-distance dispersal 1: principles of dispersal and evolution. *Q. Rev. Biol.* 41(3): 247–270.

Carr, A. F., Jr. 1940. A contribution to the herpetology of Florida. *Univ. Florida. Publ. Biol. Sci. Ser.* 3(1). 118 pp.

Carr, J. B. 1966. Permanency of the continents. *Nature* 209: 341–348.

Carter, M. R. 1989. The biology and ecology of the tiger beetles (Coleoptera: Cicindelidae) of Nebraska. *Trans. Nebr. Acad. Sci.* 17: 1–18.

Cartwright, O. L. 1935. The tiger beetles of South Carolina with the description of a new variety of *Tetracha virginica* (L.) (Coleoptera: Cicindelidae). *Bull. Brooklyn Entomol. Soc.* 30(2): 69–77.

———. 1939. Eleven new American Coleoptera. *Ann. Entomol. Soc. Am.* 32(2): 353–364.

Casey, T. L. 1897. Coleopterological notices. VII. *Ann. N.Y. Acad. Sci.* 9: 285–684.

———. 1909. Studies in the Caraboidea and Lamellicornia. *Can. Entomol.* 41(8): 253–284.

———. 1913. *Studies in the Cicindelidae and Carabidae of America.* Lancaster, Pa.: New Era. 192 pp.

———. 1914a. *Omus* and *Cicindela* keys and descriptions of North American species. *Memoirs Coleopt.* 5: 1–24.

———. 1914b. Miscellaneous notes and new species. *In: Memoirs on the Coleoptera 5*, pp. 355–378. Lancaster, Pa.: New Era.

———. 1916. *Further studies in the Cicindelidae.* Lancaster, Pa.: New Era. 34 pp.

———. 1924. *Additions to the known Coleoptera of North America.* Lancaster, Pa.: New Era. 347 pp.

Cassola, F. 1994. Studies on tiger beetles, LXXIV. Notes on the poorly known Mexican taxa of the Chaudoir's collection (Coleoptera: Cicindelidae). *Doriana* 6(283): 1–6.

Cassola, F., and H. Sawada. 1990. Two new tiger beetles from Yucatan, Mexico. (Studies on tiger beetles, 64) (Coleoptera, Cicindelidae). *Doriana* 6(272): 1–10.

Castle, D. M., and P. Laurent. 1896. April collecting in Georgia and Florida. *Entomol. News* 7(10): 300–305.

Cazier, M. A. 1937a. A new California *Omus* (Coleoptera—Cicindelidae). *Pan-Pac. Entomol.* 13: 94.

———. 1937b. Review of the *willistoni, fulgida, parowana* and *senilis* groups of the genus *Cicindela* (Coleoptera—Cicindelidae). *Bull. South. Calif. Acad. Sci.* 35(3): 156–163.

———. 1939a. Two new western tiger beetles with notes (Coleoptera—Cicindelidae). *Bull. Brooklyn Entomol. Soc.* 34: 24–48.

———. 1939b. Notes on the genus *Amblycheila*. *Pan-Pac. Entomol.* 15: 110.

———. 1942. A monographic revision of the genus *Omus* (Coleoptera—Cicindelidae). Ph.D. diss., University of California, Berkeley. 434 pp.

———. 1948. The origin, distribution, and classification of the tiger beetles of lower California (Coleoptera: Cicindelidae). *Am. Mus. Novit.* 1382: 1–27.

———. 1954. A review of the Mexican tiger beetles of the genus *Cicindela* (Coleoptera: Cicindelidae). *Bull. Am. Mus. Nat. Hist.* 103(3): 231–309.

———. 1960. Notes on Mexican tiger beetles belonging to the genus *Cicindela*. *Am. Mus. Novit.* 2025: 1–12.

Chaney, R. W. 1940. Tertiary forests and continental history. *Bull. Geol. Soc. Am.* 51: 469–488.

———. 1947. Tertiary centers and migration routes. *Ecol. Monogr.* 17: 139–148.

Chapman, R. N., C. E. Mickel, G. E. Miller, J. R. Parker, and E. G. Kelley. 1926. Studies in the ecology of sand dune insects. *Ecology* 7: 416–427, 549–557.

Chevrolat, L. A. A. 1863. Coleopteres de l'ile de Cuba. Notes, synonymies, et descriptions d'especes nouvelles. *Ann. Soc. Entomol. Fr.* 3: 183–210.

Choate, P. M. 1975. Notes on Cicindelidae in South Carolina. *Cicindela* 7(4): 71–76.

———. 1984. A new species of *Cicindela* Linnaeus (Coleoptera: Cicindelidae) from Florida, and elevation of *C. abdominalis scabrosa* Schaupp to species. *Entomol. News* 95(3): 73–82.

———. 1990. Checklist of the ground beetles of Florida (Coleoptera: Carabidae) literature records. *Fla. Entomol.* 73(3): 476–492.

Christman, R. 1953. Geology of St. Bartholomew, St. Martin, and Anguilla, Lesser Antilles. *Geol. Soc. Am. Bull.* 64: 65–96.

Ciegler, J. C. 1997. Tiger beetles of South Carolina (Coleoptera: Cicindelidae). *Coleopt. Bull.* 51: 177–192.

———. 2000. *Ground beetles and wrinkled bark beetles of South Carolina (Coleoptera: Geadephaga: Carabidae and Rhysodidae). Biota of South Carolina.* Vol. 1. Clemson, S.C.: Clemson University. 149 pp.

Clancy, P. 1996. Stalking tigers of the beach. *Nature Conservancy* 46(3): 8–9.

Clark, R. B. 1956. Species and systematics. *Syst. Zool.* 5: 1–10.

Clausen, R. T. 1941. On the use of the terms subspecies and variety. *Rhodora* 43: 157–167.

Clements, F. E., and R. W. Chaney. 1937. Environment and life in the Great Plains. *Carnegie Inst. Wash., Suppl. Publ.* 24: 1–54.

Clench, W. J. 1954. The occurrence of clines in molluscan populations. *Syst. Zool.* 3: 122–125.

Cole, G. 1984. The behavioral physiology of the escape response of tiger beetle larvae (*Cicindela*). *Int. Congr. Entomol. Proc.* 17: 232.

Colless, D. H. 1967. The phylogenetic fallacy. *Syst. Zool.* 16(4): 289–295.

———. 1969a. The phylogenetic fallacy revisited. *Syst. Zool.* 18(1): 115–126.

———. 1969b. The interpretation of Hennig's phylogenetic systematics: a reply to Dr. Schlee. *Syst. Zool.* 18(1): 134–144.

Comboni, D. J., and T. D. Schultz. 1989. New state records for two tiger beetles (Coleoptera: Cicindelidae) in southern New England. *Entomol. News* 100(4): 150–152.

Comstock, J. H. 1918. *The wings of insects.* Ithaca, N.Y.: Comstock. 430 pp.

———. 1920. *An introduction to entomology.* Ithaca, N.Y.: Comstock. 1064 pp.

Comstock, J. H., and V. L. Kellogg. 1912. *Elements of insect anatomy.* Ithaca, N.Y.: Comstock. 145 pp.

Conant, R. 1960. The queen snake, *Natrix septemvittata*, in the interior highlands of Arkansas and Missouri, with comments upon similar disjunct distributions. *Proc. Acad. Nat. Sci. Phila.* 112: 25–40.

Contreras, V. H., and B. M. Castillon. 1968. Morphology and origin of salt domes of Isthmus of Tehuantepec. *In:* J. Braunstein and G. D. O'Brien, eds., *Diapirism and diapirs. Am. Assoc. Pet. Geol. Mem.* 8: 244–260.

Cooke, C. W. 1939. Scenery of Florida as interpreted by a geologist. *Fla. Geol. Surv. Bull.* 17. 118 pp.

———. 1945. Geology of Florida. *Fla. Geol. Surv. Bull.* 29. 339 pp.

Cooke, C. W., and D. S. Mossom. 1929. Geology of Florida. *Fla. Geol. Surv. Rep.* 20: 29–227.

Cooper, K. W. 1992. An annotated, analytical bibliography of "monstrous" cicindeline beetles, and some problems they awaken. *Entomol. News* 103(1): 1–11.

Crenshaw, J. W., Jr., and W. F. Blair. 1959. Relationship in the *Pseudacris nigrita* complex in southern Georgia. *Copeia* 3: 215–222.

Cresson, E. T. 1861. Catalogue of the Cicindelidae of North America. *Proc. Am. Entomol. Soc. Phila.* 1(3): 7–21.

Criddle, N. 1907. Habits of some Manitoba "tiger beetles" (*Cicindela*). *Can. Entomol.* 39: 105–114.

———. 1910. Habits of some Manitoba tiger beetles (Cicindelidae). No. 2. *Can. Entomol.* 42: 9–15.

———. 1919. Fragments in the life-habits of Manitoba insects. *Can. Entomol.* 51: 97–101.

———. 1925. A new *Cicindela* from Alberta. *Can. Entomol.* 57: 127–128.

Croizat, L. 1958. *Panbiogeography, or an introductory synthesis of zoogeography, phytogeography, and geology; with notes on evolution, systematics, ecology, anthropology, etc.* Vol. 1. *The New World.* Caracas, Venezuela: published by the author. xxxi + 1018 pp.

———. 1962. *Space, time, form. The biological synthesis.* Caracas, Venezuela: published by the author. xix + 818 pp.

Croizat, L., G. Rosen, and G. Nelson. 1974. Centers of origin and related concepts. *Syst. Zool.* 23: 265–287.

Crombie, A. 1947. Interspecific competition. *J. Anim. Ecol.* 16: 44–73.

Crowson, R. A. 1946. Distribution and phylogeny of Cicindelidae (Coleoptera). *Entomol. Mon. Mag.* 82: 278.

———. [1954] 1967. *The natural classification of families of Coleoptera.* Reprint, London: E. W. Classey.

Cummins, M. P. 1992. Amphibious behavior of a tropical, adult tiger beetle, *Oxycheila polita* Bates (Coleoptera: Cicindelidae). *Coleopt. Bull.* 46(2): 145–151.

Cutler, B. 1973. Adult behaviour of *Cicindela cursitans*. *Cicindela* 5(4): 77–80.

———. 1982. *Phidippus pius* (Araneae: Salticidae) prey of *Cicindela fulgida fulgida* (Coleoptera: Cicindelidae). *Cicindela* 14(1–4): 34.

Dabrowski, M. J. 1971. Palynochronological materials—Eemian interglacial. *Bulletin de l'Academie Polonaise des Sciences Serie des Sciences de la Terre* 19: 29–36.

Dahl, R. G. 1939. A new California tiger beetle (Coleoptera-Cicindelidae). *Bull. Brooklyn Entomol. Soc.* 34: 221–222.

———. 1940. Notes on some Cicindelidae (Coleoptera). *Pan-Pac. Entomol.* 16: 79–80.

———. 1941. The Leng types of Cicindelidae (Coleoptera). *Entomol. News* 52: 169–172, 188–191.

Dall, W. H., and S. Brown, Jr. 1894. Cenozoic geology along the Apalachicola River. *Bull. Geol. Soc. Am.* 5: 147–170.

Darlington, P. J., Jr. 1931. The cicindelid and carabid beetles of New Hampshire, with special reference to their geographic relationships. Ph.D. diss., Harvard University, Cambridge. 164 pp.

———. 1936. Variation and atrophy of flying wings of some Carabid beetles. *Ann. Entomol. Soc. Am.* 29: 136–175.

———. 1938. The origin of the fauna of the Greater Antilles, with discussion of dispersal of animals over water and through the air. *Q. Rev. Biol.* 13: 274–300.

———. 1943. Carabidae of mountains and islands: data on the evolution of isolated faunas, and on atrophy of wings. *Ecol. Monogr.* 13: 37–61.

———. 1948. Geographic distribution of cold blooded vertebrates. *Q. Rev. Biol.* 23: 1–26, 105–123.

———. 1959a. Area, climate, and evolution. *Evolution* 13: 488–510.

———. 1959b. Darwin and zoogeography. *Proc. Am. Phil. Soc.* 103: 307–319.

———. 1964. Drifting continents and late Paleozoic geography. *Proc. Natl. Acad. Sci.* 52 (4): 1084–1091.

———. 1970a. Carabidae on tropical islands, especially the West Indies. *Biotropica* 2(1): 7–15.

———. 1970b. A practical criticism of Hennig-Brunding "phylogenetic systematics" and Antarctic biogeography. *Syst. Zool.* 19: 1–18.

———. 1971. Modern taxonomy, reality, and usefulness. *Syst. Zool.* 20(3): 341–365.

Darwin, C. [1859] 1968. *The origin of species by means of natural selection or the preservation of favoured races in the struggle for life.* Reprint, New York: Penguin.

Davis, A. C. 1928. A new *Cicindela* (Coleoptera, Cicindelidae). *Pan-Pac. Entomol.* 5: 65–66.

Davis, C. A. 1903. The Cicindelidae of Rhode island. *Entomol. News* 14(8): 270–273.

———. 1904. Checklist of Coleoptera from Rhode Island. Providence: Roger Williams Park Museum. 47 pp.

Davis, J. H. 1942. The ecology of the vegetation and topography of the Sand Keys of Florida. *Carnegie Inst. Washington Publ.* 524: 113–195.

———. 1946. The peat deposits of Florida. *Fla. Geol. Surv. Geol. Bull.* 30. 247 pp.

Davis, M. B. 1976. Pleistocene biogeography of temperate deciduous forests, pp. 13–26. *In*: R. C. West and W. G. Haag, eds., *Ecology of the Pleistocene*. Geoscience and Man, vol. 13. Baton Rouge: Louisiana State University Department of Geography and Anthropology.

Davis, W. T. 1911. Miscellaneous notes on collecting in Georgia. *J. N.Y. Entomol. Soc.* 18(2): 216–219.

———. 1912. Notes on the distribution of several species of tiger beetles. *J. N.Y. Entomol. Soc.* 20: 17–20.

———. 1916. Notes on tiger beetles from North Carolina. *J. N.Y. Entomol. Soc.* 24(2): 154–155.

———. 1918a. *Cicindela abdominalis* Fab. *J. N.Y. Entomol. Soc.* 26(1): 48.

———. 1918b. A new tiger beetle from Texas. *Bull. Brooklyn Entomol. Soc.* 12: 33–34.

———. 1921. *Cicindela tranquebarica* and its habits. *Bull. Brooklyn Entomol. Soc.* 16(5): 111.

Dawson, R. W., and W. Horn. 1928. *The tiger beetles of Minnesota.* University of Minnesota Agricultural Experiment Station, Technical Bulletin 56. 13 pp.

Day, E. 1969. Black widow spider victimizes tiger beetle. *Cicindela* 1(2): 8.

DeBach, P. 1966. The competitive displacement and coexistence principles. *Annu. Rev. Entomol.* 11: 183–212.

Deevey, E. S., Jr. 1949. Biogeography of the Pleistocene I. Europe and North America. *Bull. Geol. Soc. Am.* 60: 1315–1416.

———. 1950. Hydroids from Louisiana and Texas, with remarks on the Pleistocene biogeography of the western Gulf of Mexico. *Ecology* 31: 334–367.

Dejean, P.F.M.A. 1825. *Species general des Coleopteres de la collecion de M. le comte Dejean.* Vol. 1. Paris: Crevot. 463 pp.

Dengo, G. 1953. Geology of the Caracas region, Venezuela. *Geol. Soc. Am. Bull.* 64: 7–40.

Desender, K., L. Baert, and J. P. Maelfait. 1992. Distribution and speciation of carabid beetles in the Galapagos Archipelago (Ecuador). *Bulletin de l'Institut Royal des Sciences Naturelles de Belgique Entomologie* 62: 57–65.

Dietz, R. S., and J. C. Holden. 1970. Reconstruction of Pangea; breakup and dispersion of continents, Permian to present. *J. Geophys. Res.* 75 (26): 4939–4956.

Dillon, E. S., and L. S. Dillon. 1972. *A manual of common beetles of eastern North America.* New York: Dover. 434 pp.

Dobzhansky, T. 1950. Evolution in the tropics. *Am. Sci.* 38: 209–221.

Donnelly, J. 1986. Vermont to protect tiger beetle. *Caledonian Record*, 4 August 1986, 70(286): 2.

Dorf, E. 1959. Climatic changes of the past and present. *Contrib. Mus. Paleontol. Univ. Mich.* 13: 181–208, pl. 1.

———. 1960. Climatic changes of the past and present. *Am. Sci.* 48: 341–364.

Doubleday, E. 1838. Communications on the natural history of North America. *Entomol. Mag.* 5(5): 409–417.

Dow, R. P. 1913. The makers of Coleopterous species. *Bull. Brooklyn Entomol. Soc.* 8: 51–54.

Dowling, H. G. 1956. Geographic relationships of Ozarkian amphibians and reptiles. *Southwest. Nat.* 1: 174–189.

Downie, N. M., and R. H. Arnett, Jr. 1996. *The beetles of northeastern North*

America. Vol. 1, *Introduction; suborders Archostemata, Adephaga, and Polyphaga, thru superfamily Cantharoidea*. Gainesville, Fla.: Sandhill Crane Press. 880 pp.
Dozier, H. L. 1918. An annotated list of Gainesville, Florida Coleoptera. *Entomol. News* 29(8): 295–298.
Dreisig, H. 1978. Sand springers: predators of the sands. *Nat. Verden* 11: 337–346.
Drew, W. A., and H. W. van Cleave. 1962. The tiger beetles of Oklahoma (Cicindelidae). *Proc. Okla. Acad. Sci.* 42: 101–122.
Duncan, D. K. 1958. A new subspecies of the genus *Cicindela* (Coleoptera: Cicindelidae). *Pan-Pac. Entomol.* 34(1): 43.
Dunn, G. A. 1979. A New Hampshire population of *Cicindela ancocisconensis* exhibiting reduced elytral maculation. *Cicindela* 11(4): 61–64.
———. 1980a. A note on the night-time mating activity of *Cicindela lepida* (Coleoptera: Cicindelidae). *Gt. Lakes Entomol.* 13(3): 167.
———. 1980b. Taking *Amblychila cylindriformis* Say by barrier-type pitfall trap (Coleoptera: Cicindelidae). *Entomol. News* 91(4): 143–144.
———. 1981. The tiger beetles of New Hampshire. *Cicindela* 13(1–2): 1–28.
———. 1983. Tiger beetle wine. *Cicindela* 15(1/4): 34.
———. 1986. Tiger beetles of New England (Coleoptera: Cicindelidae). *Y.E.S. Quarterly* 3(1): 27–41.
———. 1987. Additions to the list of Carabidae and Cicindelinae (Coleoptera) of the Beaver Islands, Charlevoix Co., Michigan (USA). *Y.E.S. Quarterly* 4(2): 11.
———. 1993a. Beach tiger beetles in an onion field? *Y.E.S. Quarterly* 10(4): 34–35.
———. 1993b. Tiger beetles of Hoffmaster State Park, Muskegon County, Michigan. *Y.E.S. Quarterly* 10(2): 33–35.
———. 1994. Interesting beetle observations on Mackinac Island, Michigan. *Y.E.S. Quarterly* 11(3): 114–116.
Dunn, G. A., and D. A. Wilson. 1979. *Cicindela marginipennis* in New Hampshire. *Cicindela* 11(4): 49–56.
Dunn, G. W. 1891. Tiger beetles of California. *Zoe Biol. J.* 2: 152–154.
Durham, J. W. 1963. Palaeogeographic conclusions in light of biological data, pp. 355–365. *In: Pacific Basin Biogeography: A Symposium: Tenth Pacific Science Congress, Honolulu, Hawaii, 1961*. Honolulu: Bishop Museum Press.
Durrant, S. D. 1955. In defense of the subspecies. *Syst. Zool.* 4: 186–190.
Eckhoff, D. E. 1939. The Cicindelidae of Iowa. *Iowa State J. Sci.* 13(2): 201–230.
Edmonds, W. D. 1972. Comparative skeletal morphology, systematics, and evolution of the Phanaeine dung beetles (Coleoptera; Scarabaeidae). *Univ. Kans. Sci. Bull.* 49(11): 731–874.

Edwards, J. G. 1949. *Coleoptera or beetles east of the Great Plains.* Ann Arbor: Edwards Bros. 181 pp.

———. 1954. A new approach to infraspecific categories. *Syst. Zool.* 13: 1–20.

Elliott, N., and P. Salbert. 1978. Notes on tiger beetles *Megacephala carolina*, *Cicindela trifasciata* of San Salvador Island, Bahamas. *Cicindela* 10(2): 21–22.

Emden, F. I. van. 1935. Die larven der Cicindelinae. I. Einleitendes und alocosternale Phyle. *Tijdschrift voor Entomologie* 78: 134–183.

Emerson, A. E. 1952. The biogeography of termites. *Bull. Am. Mus. Nat. Hist.* 99: 217–225.

Emiliani, C. 1961. Isotopic paleotemperatures. *Science* 154: 851–857.

———. 1972. Quaternary hypsithermal. *Quat. Res.* 2: 270–273.

Erwin, T. L. 1970. A reclassification of the bombardier beetles and a taxonomic revision of the North and Middle American species. *Quaest. Entomol.* 6: 4–215.

———. 1979. Thoughts on the evolutionary history of ground beetles: hypotheses generated from comparative faunal analyses of lowland forest sites and tropical regions, pp. 539–592. *In: Carabid Beetles, their evolution, natural history, and classification. Proceedings of the First International Symposium on Carabidology.* The Hague: Junk.

Erwin, T. L., and G. N. House. 1978. A catalogue of the primary types of Carabidae (incl. Cicindelinae) in the collections of the United States National Museum of Natural History (USNM) (Coleoptera). *Coleopt. Bull.* 32(3): 231–255.

Evans, H. E. 1965. A description of the larva of *Methoca stygia* (Say) with notes on other Tiphiidae. *Proc. Entomol. Soc. Wash.* 67(2): 88–95.

Evans, M. E. G. 1965. The feeding method of *Cicindela hybrida* L. *Proc. R. Entomol. Soc. Lond. Ser. A Gen. Entomol.* 40(4–6): 61–66.

———. 1977. Locomotion in the Coleoptera Adephaga, especially Carabidae. *J. Zool. Lond.* 181: 189–226.

Evans, W. H. 1949. *A catalogue of the Hesperiidae from Europe, Asia and Australia in the British Museum (Natural History).* London: Printed by order of the trustees of the British Museum. 502 pp.

———. 1951. *A catalogue of the American Hesperiidae in the British Museum (Natural History).* Part 1, *Pyrrhopyginae.* London: Printed by order of the trustees of the British Museum. 92 pp.

———. 1952. *A catalogue of the American Hesperiidae in the British Museum (Natural History).* Part II. *Pyrginae.* Sect. 1. Norwich: Jarold and Sons. 178 pp.

———. 1955. *A catalogue of the American Hesperiidae in the British Museum (Natural History).* Part IV. Hesperiinae and Megathyminae. London: Printed by order of the trustees of the British Museum. 499 pp.

Ewing, J. L., J. L. Worzel, and M. Ewing. 1962. Sediments and oceanic structural history of the Gulf of Mexico. *J. Geophys. Res.* 67: 2509–2527.

Fabricius, J. C. 1801. *Systema Eleutheratorum I.* Kiliae. 506 pp.

Fackler, H. L. 1918. The tiger beetles of Kansas (family Carabidae; subfamily Cicindelinae; order Coleoptera). Master's thesis, University of Kansas, Lawrence. 51 pp.

Fattig, P. W. 1949. The Carabidae or ground beetles of Georgia. *Emory Univ. Mus. Bull.* 7: 1–62.

———. 1951. An unusual tiger beetle. *Coleopt. Bull.* 5(5/6): 72–73.

Fernald, M. L. 1931. Specific segregations and identities of some floras of eastern North America and the Old World. *Rhodora* 33: 25–83.

Ferris, C. D. 1969. Notes on collecting early *Cicindela* in eastern Wyoming. *Cicindela* 1(2): 9–13.

Fielding, K., and C. B. Knisley. 1995. Mating behavior in two tiger beetles, *Cicindela dorsalis* and *C. puritana* (Coleoptera: Cicindelidae). *Entomol. News* 106(2): 61–67.

Fink, L. K., Jr. 1970. Evidence for the antiquity of the Lesser Antilles arc. *Trans. Am. Geophys. Union* 51: 326–327.

Flint, R. F. 1940. Pleistocene features of the Atlantic Coastal plain. *Am. J. Sci.* 238: 757–787.

Forbes, W. T. M. 1922. The wing venation of the Coleoptera. *Ann. Entomol. Soc. Am.* 15: 328–345.

Forsyth, D. J. 1970. The structure of the defense glands of Cicindelidae, Amphizoidae, and Hygrobiidae (Insecta: Coleoptera). *J. Zool.* 160: 51–69.

Foster, R. F., and P. Knowles. 1990. Genetic variance in five species and 12 populations of *Cicindela* (Coleoptera: Cicindelidae) in northwestern Ontario. *Ann. Entomol. Soc. Am.* 83(4): 838–845.

Fowler, H. G. 1987. Predatory behavior of *Megacephala fulgida* (Coleoptera: Cicindelidae). *Coleopt. Bull.* 41(4): 407–408.

Fox, H. 1910. Observations on Cicindelidae in Northern Cape May County, N. J., during the summers of 1908–1909. *Entomol. News* 21: 75–82.

Fox, R. M. 1955. On subspecies. *Syst. Zool.* 4: 93–95.

Franklin, R. T. 1988. *Cicindela unipunctata* from pitfall traps (Coleoptera: Cicindelidae). *J. Kans. Entomol. Soc.* 61(2): 249–250.

Freeland, O. L., and R. S. Dietz. 1972. Plate tectonics in the Caribbean: a reply. *Nature (Lond.)* 235: 156–157.

Freitag, R. 1965. A revision of the North American species of the *Cicindela maritima* group with a study of hybridization between *Cicindela duodecimguttata* and *oregona*. *Quaest. Entomol.* 1: 87–170.

———. 1966. The female genitalia of four species of tiger beetles. *Can. Entomol.* 98: 942–952.

———. 1969. A revision of the species of the genus *Evarthrus* LeConte. *Quaest. Entomol.* 5: 89–21.

———. 1972. Female genitalia of the North American species of the *Cicindela maritima* group (Coleoptera: Cicindelidae). *Can. Entomol.* 104(8): 1277–1306.

———. 1974. Selection for a non-genitalic mating structure in female tiger beetles of the genus *Cicindela* (Coleoptera: Cicindelidae). *Can. Entomol.* 106(6): 561–568.

———. 1975. An unusual structure in females of *Cicindela albicans* Chaud. and related species. *J. Aust. Entomol. Soc.* 14: 319–320.

———. 1979. Reclassification, phylogeny, and zoogeography of the Australian species of *Cicindela*. *Aust. J. Zool. Suppl. Ser.* 66: 1–99.

———. 1985. Additional Jamaican records for *Cicindela carthagena* Dejean. *Cicindela* 17(2): 35–36.

———. 1992. Biogeography of West Indian tiger beetles (Coleoptera: Cicindelidae), pp. 123–158. *In:* G. R. Noonan, G. E. Ball, and N. E. Stork, eds., *The biogeography of ground beetles of mountains and islands.* Andover, Hampshire: Intercept.

———. 1999. *Catalogue of the tiger beetles of Canada and the United States.* Ottawa: NRC Research Press. 195 pp.

Freitag, R., and B. L. Barnes. 1989. Classification of Brazilian species of *Cicindela* and phylogeny and biogeography of subgenera *Brasiella*, *Gaymara* new subgenus, *Plectographa* and South American species of *Cylindera* (Coleoptera: Cicindelidae). *Quaest. Entomol.* 25(3): 241–386.

Freitag, R., D. H. Kavanaugh, and R. Morgan. 1993. A new species of *Cicindela* (Coleoptera: Carabidae: Cicindelini) from remnant native grassland in Santa Cruz County, California. *Coleopt. Bull.* 47(2): 113–120.

Freitag, R., and S. K. Lee. 1972. Sound producing structures in adult *Cicindela tranquebarica* (Coleoptera: Cicindelidae) including a list of tiger beetles and ground beetles with flight wing files. *Can. Entomol.* 104(6): 851–857.

Freitag, R., J. E. Olynyk, and B. L. Barnes. 1980. Mating behavior and genitalic counterparts in tiger beetles (Carabidae: Cicindelinae). *Int. J. Invertebr. Reprod.* 2(2): 131–135.

Freitag, R., and D. L. Pearson. 1973. A note on geographical variation in *Cicindela depressula*. *Cicindela* 5(2): 29–31.

Freitag, R., L. A. Schincariol, and B. L. Barnes. 1985. A review of nomenclature for genitalic structures of *Cicindela*. *Cicindela* 17(2): 17–27.

Freitag, R., and R. Tropea. 1969. Twenty-one Cicindelid species in thirty-eight days. *Cicindela* 1(4): 14–27.

Frick, K. E. 1957. Biology and control of tiger beetles in alkali bee nesting sites. *J. Econ. Entomol.* 50: 503–504.

Friederichs, H. F. 1931. Beitrage zur Morphologie und Physiologie der Sehrorgane der Cicindelinen (Col.). *Z. Morphol. Oekol. Tiere* 21: 1–72.

Frye, J. C. 1973. Pleistocene succession of the central interior United States. *J. Quat. Res.* 3(2): 275–283.

Fuller, B. W., and T. E. Reagan. 1988. Comparative predation of the sugarcane borer (Lepidoptera: Pyralidae) on sweet sorghum and sugarcane. *J. Econ. Entomol.* 81(2): 713–717.

Gage, E. V. 1988. A new subspecies of *Cicindela politula* from New Mexico and a range extension for *Cicindela politula barbaranae* (Coleoptera: Cicindelidae). *Entomol. News* 99(3): 143–147.

Gage, E. V., and W. D. Sumlin III. 1986. Notes on *Cicindela nigrocoerulea subtropica* in Texas (Coleoptera: Cicindelidae). *Entomol. News* 97(5): 203–207.

Galian, J., J. Serrano, P. de la Rua, and E. J. C. Petitpierre. 1995. Localization and activity of rDNA genes in tiger beetles (Coleoptera: Cicindelinae). *Heredity* 74(5): 524–530.

Galian, J., A. S. Ortiz, and J. Serrano. 1990. Karyotypes of nine species of Cicindelini and cytotaxonomic notes on Cicindelinae (Coleoptera, Carabidae). *Genetica* 82: 17–24.

Gates, R. R. 1938. The species concept in the light of cytology and genetics. *Am. Nat.* 72: 340–349.

———. 1951. The taxonomic units in relation to cytogenetics and gene ecology. *Am. Nat.* 85: 31–50.

Gaumer, G. C. 1969a. Coastal tiger beetles of Texas in the genus *Cicindela* (Coleoptera: Cicindelidae). *Cicindela* 1: 2–16.

———. 1969b. A note on the activity of *Amblycheila cylindriformis*. *Cicindela* 1(1): 24.

———. 1970. Collecting *Cicindela depressula eureka* with notes on its habitat and variation. *Cicindela* 2(3): 14–17.

———. 1973. Aestival tiger beetle fauna of Big Bend National Park. *Cicindela* 5: 13–19.

———. 1977. The variation and taxonomy of *Cicindela formosa* Say (Coleoptera: Cicindelidae). Ph.D. diss., Texas A&M University, College Station. 253 pp.

Gaumer, G. C., E. J. Kurczewski, and F. E. Kurczewski. 1970. The tiger beetles of Presque Isle State Park, Pennsylvania. *Cicindela* 2(2): 4–7.

Gaumer, G. C., and R. R. Murray. 1971. Checklist of the Cicindelidae of Texas with regional distributions. *Cicindela* 3(1): 9–12.

Gilbert, C. 1984. Visual processing and the control of prey localization and pursuit behavior of the tiger beetle. *Int. Congr. Entomol. Proc.* 17: 195.

———. 1986a. Behavioral studies of visual movement perception by larval and adult tiger beetles (Cicindelidae). Dissertation Abstracts International, Section B, Sciences and Engineering 47(6): 2288.

———. 1986b. A morphological and cinematographic analysis of tiger beetle predatory behaviour (Carabidae: Cicindelinae), pp. 43–57. *In*: Boer, P. J. den, M. L. Luff, D. Mossakowski, and F. Weber, eds., *Carabid beetles, their adaptations and dynamics*. Stuttgart: Gustav Fischer.

———. 1989. Visual determinants of escape in tiger beetle larvae (Cicindelidae). *J. Insect Behav.* 2(4): 557–574.

Gilbertson, G. I. 1929. The Cicindelidae of South Dakota. *Proc. S.D. Acad. Sci.* 29: 22–26.

Gillham, N. W. 1956. Geographic variation and subspecies concept in butterflies. *Syst. Zool.* 5: 110–120.

Gilyarov, M. S., and I. K. H. Sharova. 1954. Larvae of tiger beetles (Cicindelidae) (in Russian; English summary). *Zool. Zh.* 33: 598–615.

Gissler, C. 1879. The anatomy of *Amblycheila cylindriformis* Say. *Psyche* 2: 233–248.

Glaser, J. D. 1976. Cicindelids of Chesapeake Bay revisted. *Cicindela* 8(1): 17–20.

———. 1977. Letter re: decimation of *C. dorsalis* population on Maryland portion of Assateague Is. *Cicindela* 9(1): 12.

———. 1984. The Cicindelidae (Coleoptera) of Maryland. *Md. Entomol.* 2(4): 65–76.

———. 1992. *Cicindela ancocisconensis* Harris (Coleoptera: Cicindelidae) in Maryland. *Md. Entomol.* 3(4): 145–146.

Gleadall, I. G., T. Hariyama, and Y. Tsukahara. 1989. The visual pigment chromophores in the retina of insect compound eyes, with special reference to the Coleoptera. *J. Insect Physiol.* 35(10): 787–795.

Goldschmidt, R. 1935. Geographische variation und Artbildung. *Naturwissenschaften* 23: 169–176.

Goldsmith, W. M. 1916. Field notes on the distribution and life habits of the tiger beetles (Cicindelidae) of Indiana. *Proc. Indiana Acad. Sci.* 26: 447–455.

———. 1919. A comparative study of the chromosomes of the tiger beetles (Cicindelidae). *J. Morphol.* 32: 437–489.

Gould, A. A. 1834. On the cicindelidae of Massachusetts. *Boston J. Nat. Hist.* 1: 41–54.

Gould, H. R., and R. H. Stewart. 1955. Continental terrace sediments in the northeastern Gulf of Mexico; finding ancient shorelines; a symposium with discussion. *Soc. Econ. Palaeontologists and Mineralogists*: 2–21.

Graham, S. A. 1922. A study of the wing venation of the Coleoptera. *Ann. Entomol. Soc. Am.* 15: 191–201.

Graves, R. C. 1962. Predation on *Cicindela* by a dragonfly. *Can. Entomol.* 94(11): 1231.

———. 1963. The Cicindelidae of Michigan (Coleoptera). *Am. Midl. Nat.* 69(2): 492–507.

———. 1965. The distribution of tiger beetles in Ontario (Coleoptera: Cicindelidae). *Proc. Entomol. Soc. Ontario* 95: 63–70.

———. 1969. An upper Michigan population of *Cicindela repanda* with reduced elytral maculae (Coleoptera: Cicindelidae). *Coleopt. Bull.* 23: 86–88.

———. 1981. Offshore flight in *Cicindela trifasciata*. *Cicindela* 13(3–4): 45–46.

———. 1982. Another record of offshore flight in *Cicindela trifasciata*. *Cicindela* 14(1/4): 18.

———. 1988. Geographic distribution of the North American tiger beetle *Cicindela hirticollis* Say. *Cicindela* 20(1): 1–21.

Graves, R. C., and D. W. Brzoska. 1991. The tiger beetles of Ohio (Coleoptera: Cicindelidae). *Bull. Ohio Biol. Surv.* 8(4). vi + 42 pp.

Graves, R. C., M. E. Krejci, and A. C. F. Graves. 1988. Geographic variation in the North America tiger beetle, *Cicindela hirticollis* Say, with a description of five new subspecies (Coleoptera: Cicindelidae). *Can. Entomol.* 120(7): 647–678.

Graves, R. C., and D. L. Pearson. 1973. The tiger beetles of Arkansas, Louisiana, and Mississippi (Coleoptera: Cicindelidae). *Trans. Am. Entomol. Soc.* 99(2): 157–203.

Griffiths, G. C. D. 1976. The future of Linnaean nomenclature. *Syst. Zool.* 25(2): 168–173.

Grodnitsky, D. L., and P. P. Morozov. 1995. The vortex wakes of flying beetles. *Zool. Zh.* 74(3): 66–72.

Guérin-Méneville, F. E. 1840. Cicindeles nouv. decouv a Pensacola. *Rev. Zool. (Paris)* 3: 37.

Guerra, J. F. 1993. Some observations of the termite mound-dwelling tiger beetle, *Cheilonycha auripennis* Lucas, from northeastern Bolivia. *Cicindela* 25(1/2): 23–26.

Guido, A. S., and H. G. Fowler. 1988. *Megacephala fulgida* (Coleoptera: Cicindelidae): a phonotactically orienting predator of *Scapteriscus* mole crickets (Orthoptera: Gryllotalpidae). *Cicindela* 20(3–4): 51–52.

Gundlach, J. 1891. *Contribución a la entomología cubana.* Parte quinta, Coleópteros. Vol. 3, pp. 1–404. Havana: G. Montiel.

Gunter, H. 1948. Elevations in Florida. *Fla. Geol. Surv. Geol. Bull.* 32: 1–1158.

Hadley, N. F., C. B. Knisley, T. D. Schultz, and D. L. Pearson. 1990. Water relations of tiger beetle larvae (*Cicindela marutha*): correlations with habitat microclimate and burrowing activity. *J. Arid Environ.* 19(2): 189–197.

Hairston, N. G., and C. H. Pope. 1948. Geographic variation and speciation in Appalachian salamanders (*Plethodon jordani* group). *Evolution* 2(3): 266–278.

Halffter, G. 1965. Algunas ideas acerca de la zoogeographica de America. *Rev. Soc. Mex. Hist. Nat.* 26: 1–16.

———. 1974. Elements anciens de l'entomofaune neotropicale: ses implications biogeographiques. *Quaest. Entomol.* 10: 223–262.

Hamilton, C. C. 1925. Studies on the morphology, taxonomy and ecology of the larvae of holarctic tiger beetles (family Cicindelidae). *Proc. U.S. Natl. Mus.* 65: 1–87.

Harland, S. C. 1936. The genetical concept of the species. *Biol. Rev.* 11: 83–112.

Harper, R. M. 1921. Geography of central Florida. Florida Geological Survey 7th Annual Report, pp. 71–307.

Harris, E. D. 1901. Cicindelidae of Mt. Desert, Maine. *J. N.Y. Entomol. Soc.* 9(1): 27–28.

———. 1902. Notes on Cicindelae in North Carolina. *Can. Entomol.* 34(8): 217–218.

———. 1910. Note. Cicindelidae of southern Alabama received from Löding. *J. N.Y. Entomol. Soc.* 18(2): 131.

———. 1911. *North American Cicindelidae in the Harris collection.* Yonkers, N.Y.: Truan Press. 68 pp.

———. 1913. Three new cicindelids. *J. N.Y. Entomol. Soc.* 21: 67–69.

———. 1917. Some White Mountain Cicindelae. *J. N.Y. Entomol. Soc.* 25(2): 136–138.

———. 1918. Cicindelidae of New Hampshire. *J. N.Y. Entomol. Soc.* 26(3/4): 237–238.

Harris, E. D., and C. W. Leng. 1916. *The Cicindelidae of North America as arranged by Dr. Walther Horn in Genera Insectorum.* New York: American Museum of Natural History. vi + 23 pp.

Harrison, C. G., and M. M. Ball. 1973. The role of fracture zones in a sea floor spreading. *J. Geophys. Res.* 78: 7776–7785.

Hatch, M. H. 1938. Coleoptera of Washington: Carabidae: Cicindelinae. *Univ. Wash. Publ. Biol.* 1(5): 225–240.

———. 1953. *The beetles of the Pacific Northwest.* Part 1, *Introduction and Adephaga.* Seattle: University of Washington Press. 340 pp.

Hawkeswood, T. J. 1992. A list and notes on some nocturnally active beetles (Coleoptera) attracted to street lights at Townsville, north-eastern Queensland, Australia. *G. Ital. Entomol.* 6(30): 5–8.

Hays, J. D., and A. Perruzza. 1972. The significance of calcium carbonate oscillations in eastern equatorial Atlantic deep-sea sediments for the end of the Holocene warm interval. *Quat. Res.* 2: 355–362.

Hays, J. D., T. Saito, H. D. Updyke, and L.H. Burckle. 1969. Pliocene-Pleistocene sediments of the equatorial Pacific. Their paleomagnetic, biostratigraphic, and climatic record. *Geol. Soc. Am. Bull.* 80: 1481–1514.

Hennig, W. 1965. Phylogenetic systematics. *Annu. Rev. Entomol.* 10: 97–116.

———. 1966. *Phylogenetic systematics.* Urbana: University of Illinois Press. 263 pp.

Hentz, N. M. 1830. Description of eleven new species of North American insects. *Trans. Am. Philos. Soc.* ser. 2 3: 253–258.

Higley, L. G. 1986. Morphology of reproductive structures in *Cicindela repanda* (Coleoptera: Cicindelidae). *J. Kans. Entomol. Soc.* 59(2): 303–308.

Hilchie, G. J. 1985. The tiger beetles of Alberta (Coleoptera: Carabidae, Cicindelini). *Quaest. Entomol.* 21(3): 319–347.

Hill, J. M., and C. B. Knisley. 1992. Frugivory in the tiger beetle, *Cicindela repanda* (Coleoptera: Cicindelidae). *Coleopt. Bull.* 46(3): 306–310.

Hoffmeister, J. E., and H. G. Multer. 1968. Geology and origin of the Florida Keys. *Geol. Soc. Am. Bull.* 79: 1487–1501.

Holeski, P. M., and R. C. Graves. 1978. An analysis of the shore beetle communities of some channelized streams in northwest Ohio (Coleoptera). *Gt. Lakes Entomol.* 11: 23–26.

Holman, J. A. 1965 (1966). A huge Pleistocene box turtle from Texas. *Q. J. Fla. Acad. Sci.* 28(4): 345–348.

———. 1965a. Pleistocene snakes from the Seymour foundation of Texas. *Copeia* 1: 102–104.

———. 1965b. A small Pleistocene herpetofauna from Houston, Texas. *Tex. J. Sci.* 17 (4): 418–423.

———. 1966. The Pleistocene herpetofauna of Miller's Cave, Texas. *Tex. J. Sci.* 18(4): 372–377.

Hood, L. E. 1903. Notes on *Cicindela hentzi*. *Entomol. News* 14: 113–116.

Hooper, R. R. 1969. A review of Saskatchewan tiger beetles. *Cicindela* 1(4): 1–4.

Hopkins, D. M., and J. V. Matthews. 1971. A Pliocene flora and insect fauna from the Bering Strait region. *Palaeogeogr. Palaeoclimatol. Palaeoecol.* 9: 211–231.

Horn, G. H. 1868. The United States species of *Cicindela*. *Trans. Am. Entomol. Soc.* 1: 2–3.

———. 1876. The sexual characters of North American Cicindelidae, with notes on some groups of *Cicindela*. *Trans. Am. Entomol. Soc.* 5: 232–240.

———. 1878. Descriptions of the larvae of the North American genera of Cicindelidae, also of *Dicaelus*, with a note on Rhynchophorus. *Trans. Am. Entomol. Soc.* 7: 28–40.

Horn, H. 1966. Measurement of overlap in comparative ecological studies. *Am. Nat.* 100: 419–424.

Horn, W. 1903. List of the Cicindelidae of Mexico and their relationship with the species of the United States. *J. N.Y. Entomol. Soc.* 11(3): 213–221.

———. 1906. Über das Vorkommen von *Tetracha carolina* L. en preussischen Bernstein und die Phylogenie der *Cicindela* arten. *Dtsch. Entomol. Z.* 50: 329–336.

———. 1907. *Megacephala—Tetracha*. *Dtsch. Entomol. Z.* 51: 263–271.

———. 1908a. The larvae of *Amblychila* and *Omus*. *Dtsch. Entomol. Z.* 52: 285–286.

———. 1908b. *Genera Insectorum*, fasc. 82. 486 pp., 32 pls.

———. 1923. Studien uber neue und alte Cicindeliden. *Zool. Meded. (Leiden)* 7: 90–112.

———. 1926. Carabidae: Cicindelinae. In: *Catalogus Coleopterorum auspiciis et auxilio W. Junk editus a S. Schenkling*, vol. 1, pt. 86, 1–345. Berlin: W. Junk.

———. 1928. Notes on the tiger beetles of Minnesota. Minnesota Agricultural Experiment Station Technical Bulletin 56: 9–13.

———. 1930. Notes on the races of *Omus californicus* and a list of the Cicindelidae of America north of Mexico (Coleoptera). *Trans. Am. Entomol. Soc.* 56: 73–86.

———. 1935. On some Cicindelidae from the Pacific Coast of Mexico, the West Indies and United States. *Pan-Pac. Entomol.* 11(2): 65–66.

———. 1938. 2000 Zeichnungen von Cicindeliden. *Entomologische Beihefte aus Berlin-Dahlem.* 5: 1–71, 90 plates.

Hough Goldstein, J. A., G. E. Heimpel, H. E. Bechmann, and C. E. Mason. 1993. Arthropod natural enemies of the Colorado potato beetle. *Crop Prot.* 12(5): 334–324.

Howden, H. F. 1961. New species and a new genus of Melolonthinae from the southeastern United States. *Can. Entomol.* 93: 807–812.

———. 1963. Speculations on some beetles, barriers, and climates during the Pleistocene and pre-Pleistocene periods in some non-glaciated portions of North America. *Syst. Zool.* 12(3): 178–201.

Howden, H. F. 1966. Some possible effects of the Pleistocene on the distributions of North American Scarabaeidae. *Can. Entomol.* 98: 1177–1190.

———. 1969. Effects of the Pleistocene on North American insects. *Annu. Rev. Entomol.* 14: 39–56.

———. 1970. First South Dakota records for *Amblychila cylindriformis*. *Cicindela* 2(3): 8.

Hubbell, T. H. 1954. The naming of geographically variant populations. *Syst. Zool.* 3: 113–121.

———. 1956. Some aspects of geographic variation in insects. *Annu. Rev. Entomol.* 1: 71–88.

———. 1960. Endemism and speciation in relation to Pleistocene changes in Florida and the southeastern coastal plain. *Int. Congr. Entomol. Proc.* 11: 466–469.

Hubbell, T. H., A. M. Laessle, and J. C. Dickinson. 1956. The Flint-Chattahoochee-Apalachicola region and its environments. *Bull. Fla. State Mus.* 1(1): 1–72.

Hubbs, C. L. 1943. Criteria for subspecies, species, and genera as determined by researches on fishes. *Ann. N.Y. Acad. Sci.* 44(2): 109–121.

Huber, R. L. 1966. The Coleoptera of Minnesota. Part 1: Cicindelidae (tiger beetles). *Newsletter of the Association of Minnesota Entomologists* 1(1): 21–22.

———. 1969. More additions to the list of *opera citata*. *Cicindela* 1(3): 16, 26.

———. 1970. Taxonomic pitfalls in the Nearctic *Cicindela* (Coleoptera: Cicindelidae). *Proc. North Cent. Branch Entomol. Soc. Am.*, 92–93.

———. 1977. First Wyoming record for *Ambylycheila cylindriformis* (Say) with further distributional data. *Cicindela* 9(4): 75–76.

———. 1980a. Northerly record for *Cicindela scutellaris lecontei* in Ontario. *Cicindela* 12(2): 24.

———. 1980b. Tiger beetle *Cicindela* caught in spider Theridiidae web. *Cicindela* 12(2): 32.

———. 1986. Citational enhancements for the Boyd checklist of North American Cicindelidae. *Cicindela* 18(4): 53–57.

———. 1994. A new species of *Tetracha* from the west coast of Venezuela, with comments on genus-level nomenclature (Coleoptera: Cicindelidae). *Cicindela* 26(3–4): 49–75.

Hunt, C. B. 1967. *Physiography of the United States.* San Francisco: W. H. Freeman. 480 pp.

Hurlbert, S. 1971. The nonconcept of species diversity: a critique and alternative parameters. *Ecology* 52: 577–586.

Hurley, P. M. 1968. The confirmation of continental drift. *Sci. Am.* 218(4): 52–64.

Hutchings, J. 1987. On collecting tiger beetles (Coleoptera: Cicindelidae or Carabidae) in Trinidad and Venezuela, June 1986. *Y.E.S. Quarterly* 4(3): 59–62.

Iablokoff-Khnzorian, S. M. 1974. Remarques sur les genitalia femelles des Coleopteres et leur armure. *Ann. Soc. Entomol. Fr.* 10(2): 467–486.

Ideker, J. 1977. Field separation of *Cicindela* species by escape behavior. *Cicindela* 9(2): 39–40.

———. 1980. Predator and heat avoidance behavior of *Cicindela ocellata rectilatera*. *Cicindela* 12(3): 43–45.

Illies, J. 1965. Phylogeny and zoogeography of the Plecoptera. *Annu. Rev. Entomol.* 10: 117–140.

Imlay, R. W. 1943. Jurassic formations of Gulf Region. *Am. Assoc. Pet. Geol. Bull.* 27: 1407–1533.

———. 1944. Correlation of the Cretaceous formations of the Greater Antilles, Central America, and Mexico. *Geol. Soc. Am. Bull.* 55: 1005–1046.

Ingram, W. M. 1934. Field notes on five species of the genus *Cicindela* of the family Cicindelidae from Balboa Bay, Orange County, California. *J. Entomol. Zool.* 26: 51–52.

International Trust for Zoological Nomenclature. 1961. *International code of zoological nomenclature adopted by the XV International Congress of Entomology.* London. xviii + 176 pp.

Ivie, M. A. 1983. The Cicindelidae (Coleoptera) of the Virgin Islands. *Fla. Entomol.* 66(1): 191–199.

Jeannel, R., and E. Rivalier. 1957. Coleopteres Carabiques. *Memoires de l'Institut de Recherche Scientifique de Madagascar Serie A Biologie Animale* 8: 119–129.

Johnson, C. 1972. An analysis of geographical variation in the damselfly, *Argia apicalis* (Say). *Can. Entomol.* 104: 1515–1527.

———. 1973. Distributional patterns and their interpretation in *Hetaerina* (Odonata: Calopterygidae). *Fla. Entomol.* 56(1): 24–42.

Johnson, P. T., and J. N. Layne. 1961. A new species of *Polygenis* Jordan from

Florida, with remarks on its host relationships and zoogeographic significance. *Proc. Entomol. Soc. Wash.* 63: 115–123.

Johnson, W. N. 1972. "Hit-and-run" collecting across the Gulf states. *Cicindela* 4(1): 19–20.

———. 1979. Sympatric population of *Cicindela limbalis transversa* and *Cicindela splendida cyanocephalata*. *Cicindela* 11(2): 26.

———. 1990a. A new subspecies of *Cicindela limbata* Say from Labrador (Coleoptera: Cicindelidae). *Le Naturaliste Canadien* 116(4): 261–266.

———. 1990b. A new subspecies of *Cicindela patruela* from west-central Wisconsin. *Cicindela* 21(2): 27–32.

———. 1990c. A new subspecies of *Cicindela pusilla* Say from northern Arizona. *Cicindela* 22(1): 1–12.

———. 1991a. A new species of *Dromochorus* from southern Texas (Coleoptera: Cicindelidae). *Cicindela* 23(2/3): 49–54.

———. 1991b. Two new species of *Megacephala* from Panama (Coleoptera: Cicindelidae). *Cicindela* 23(1): 1–10.

———. 1993a. A new species of *Megacephala* from Nicaragua (Coleoptera: Cicindelidae). *Cicindela* 25(1–2): 13–22.

———. 1993b. Notes on *Cicindela yucatana* W. Horn. *Cicindela* 25(3–4): 41–44.

———. 1994. A new species of *Cicindela* from Oaxaca, Mexico (Coleoptera: Cicindelidae). *Cicindela* 26(3–4): 41–46.

Jones, S. B. 1972. A systematic study of the Fasciculate group of *Veronina*. *Brittonia* 24: 28–45.

Jonge Poerink, W. H. 1953. Caribbean tiger beetles of the genus *Cicindela*. *Studies on the fauna of Curaçao and other Caribbean islands* 4(19): 120–143.

Jordan, D. S. 1905. The origin of species through isolation. *Science* New Series 22: 545–562.

Kaulbars, M. M., and R. Freitag. 1993a. A description of the third instar larva of *Cicindela denikei* Brown. *Cicindela* 25(3–4): 45–48.

———. 1993b. Foraging behaviour of the tiger beetle *Cicindela denikei* Brown (Coleoptera: Cicindelidae). *Can. Field-Nat.* 107(1): 53–58.

———. 1993c. Geographical variation, classification, reconstructed phylogeny, and geographical history of the *Cicindela sexguttata* group (Coleoptera, Cicindelidae). *Can. Entomol.* 125(2): 267–316.

Kavanaugh, D. H. 1972. Hennig's principles and methods of phylogenetic systematics. *Biologist* 54(3): 115–127.

Keeton, W. T. 1959. A new family for the diplopod genus *Floridobolus* (Spirobolida: Spirobolidea). *Bull. Brooklyn Entomol. Soc.* 54: 1–7.

Key, K. H. L. 1968. The concept of stasipatric speciation. *Syst. Zool.* 17: 14–22.

Khudoley, K. M. 1967. Principal features of Cuban geology. *Am. Assoc. Pet. Geol. Bull.* 51: 668–677.

King, B. L. 1988. Tiger beetles of the Leaf River drainage system of southeastern Mississippi (Coleoptera: Cicindelidae). *Cicindela* 20(3–4): 41–49.

Kippenhan, M. G. 1990a. A survey of the tiger beetles (Coleoptera: Cicindelidae) of Colorado. *Entomol. News* 101(5): 307–315.

———. 1990b. Tiger beetles and ants. *Cicindela* 22(4): 53–59.

———. 1991. Notes on elytral deformities and predator damage of *Cicindela*. *Cicindela* 23(1): 11–16.

———. 1994. The tiger beetles (Coleoptera: Cicindelidae) of Colorado. *Trans. Am. Entomol. Soc.* 120(1): 1–86.

Kippenhan, M. G., and M. R. Carter. 1993. Notes on early seasonal activity of three species of *Cicindela* in Colorado. *Cicindela* 25(3–4): 49–51.

Kiriakoff, S. G. 1947. Le cline, une nouvelle categorie systematique intraspecifique. *Bull. Ann. Soc. R. Entomol. Belgique* 83: 130–140.

———. 1959. Phylogenetic systematics versus typology. *Syst. Zool.* 8(2): 117–118.

Kirk, V. M. 1969. A list of the beetles of South Carolina. Part 1, Northern coastal plain. South Carolina Agricultural Experiment Station, Technical Bulletin 1033. 124 pp.

———. 1970. A list of the beetles of South Carolina. Part 2, Mountain, Piedmont, and Southern coastal plain. South Carolina Agricultural Experiment Station, Bulletin 1038. 117 pp.

Kirk, V. M., and E. U. Balsbaugh, Jr. 1975. A list of the beetles of South Dakota. South Dakota State University, Agricultural Experiment Station, Technical Bulletin 42. 139 pp.

Kirkland, D. W., and J. E. Gerhard. 1971. Jurassic salt, Gulf of Mexico, and its temporal relation to circum-Gulf evaporites. *Am. Assoc. Pet. Geol. Bull.* 55: 680–686.

Knaus, W. 1900. The Cicindelidae of Kansas. *Can. Entomol.* 32: 109–116.

———. 1901. Collecting notes on Kansas Coleoptera. II. *Can. Entomol.* 33: 110–115.

———. 1922. Two new forms of *Cicindela* with remarks on other forms. *J. N.Y. Entomol. Soc.* 30: 194–197.

Knisley, C. B. 1978. Collecting *Cicindela willistoni estancia* beneath alkali encrustations. *Cicindela* 10(2): 31–32.

———. 1979. Distribution, abundance and seasonality of tiger beetles (Cicindelidae) in the Indiana Dunes region. *Proc. Indiana Acad. Sci.* 88: 208–217.

———. 1984. Ecological distribution of tiger beetles (Coleoptera: Cicindelidae) in Colfax County, New Mexico. *Southwest. Nat.* 29(1): 93–104.

———. 1985a. Associations of mites and tiger beetles (Coleoptera: Cicindelidae) in southeastern Arizona. *Proc. Entomol. Soc. Wash.* 87(2): 335–340.

———. 1985b. Utilization of tiger beetle larval burrows by a nest-provisioning wasp, *Leucodynerus russatus* (Bohart) (Hymenoptera: Eumenidae). *Proc. Entomol. Soc. Wash.* 87(2): 481.

———. 1987. Habitats, food resources, and natural enemies of a community of larval *Cicindela* in southeastern Arizona (Coleoptera: Cicindelidae). *Can. J. Zool.* 65(5): 1191–1200.

———. 1991. Tiger beetles. Class Insecta: order Coleoptera: family Cicindelidae: subphylum Tracheata, pp. 231–237. *In:* K. Terwilliger, ed., *Virginia's endangered species.* Blacksburg, Va.: McDonald and Woodward. 672 pp.

Knisley, C. B., D. W. Brzoska, and J. R. Schrock. 1987. Distribution, checklist and key to adult tiger beetles (Coleoptera: Cicindelidae) of Indiana. *Proc. Indiana Acad. Sci.* 97: 279–294.

Knisley, C. B., and J. M. Hill. 1992. Effects of habitat change from ecological succession and human impact on tiger beetles. *Va. J. Sci.* 43(1B): 133–142.

Knisley, C. B., and W. H. Hoback. 1994. Nocturnal roosting of *Odontocheila confusa* Dejean in the Peruvian Amazon (Coleoptera: Carabidae: Cicindelinae). *Coleopt. Bull.* 48(4): 353–354.

Knisley, C. B., and S. A. Juliano. 1988. Survival, development, and size of larval tiger beetles: effects of food and water. *Ecology* 69(6): 1983–1992.

Knisley, C. B., J. I. Luebke, and D. R. Beatty. 1987. Natural history and population decline of the coastal tiger beetle, *Cicindela dorsalis dorsalis* Say (Coleoptera: Cicindelidae). *Va. J. Sci.* 38(4): 293–303.

Knisley, C. B., and D. L. Pearson. 1981. The function of turret building behaviour in the larval tiger beetle, *Cicindela willistoni* (Coleoptera: Cicindelidae). *Ecol. Entomol.* 6(4): 401–410.

———. 1984. Biosystematics of larval tiger beetles of the Sulphur Springs Valley, Arizona. Descriptions of species and a review of larval characters for *Cicindela* (Coleoptera: Cicindelidae). *Trans. Am. Entomol. Soc.* 110(4): 465–551.

Knisley, C. B., D. L. Reeves, and G. T. Stephens. 1989. Behavior and development of the wasp *Pterombrus rufiventris hyalinatus* Krombein (Hymenoptera: Tiphiidae), a parasite of larval tiger beetles (Coleoptera: Cicindelidae). *Proc. Entomol. Soc. Wash.* 91(2): 179–184.

Knisley, C. B., and T. D. Schultz. 1997. *The biology of tiger beetles and a guide to the species of the South Atlantic States.* Martinsville: Virginia Museum of Natural History Special Publication 5. viii + 210 pp.

Knisley, C. B., T. D. Schultz, and T. H. Hasewinkel. 1990. Seasonal activity and thermoregulatory behavior of *Cicindela patruela* (Coleoptera: Cicindelidae). *Ann. Entomol. Soc. Am.* 83(5): 911–915.

Knudsen, J. W. 1985. A brief review of *Cicindela fulgida* with descriptions of three new subspecies from New Mexico (Coleoptera: Cicindelidae). *Entomol. News* 96(5): 177–187.

Koebele, A. 1878. (no title: collecting notes on Florida insects). *Bull. Brooklyn Entomol. Soc.* 1: 44.

Kolbe, H. J. 1912. Die differenzierung der zoogeographischen elemente der continente. *2nd Int. Congr. Entomol.* 433–476.

Koopman, K. F. 1958. Land bridges and ecology in bat distribution on islands off the northern coast of South America. *Evolution* 12: 429–439.

Kraus, W., and R. C. Lederhouse. 1983. Contact guarding during courtship in the tiger beetle *Cicindela marutha* Dow (Coleoptera: Cicindelidae). *Am. Midl. Nat.* 110(1): 208–211.

Krekler, C. M. 1959. Dispersal of cavernicolous beetles. *Syst. Zool.* 8: 119–130.

Kritsky, G., and S. Simon. 1995. Mandibular sexual dimorphism in *Cicindela* Linnaeus (Coleoptera: Cicindelidae). *Coleopt. Bull.* 49(2): 143–148.

Krombein, K. V. 1979. Studies in the Tiphiidae, XII. A new genus of Methocinae with notes on the subgenera of *Methoca* Latreille (Hymenoptera: Aculeata). *Proc. Entomol. Soc. Wash.* 81(3): 424–434.

Krombein, K. V., and H. E. Evans. 1976. Three new neotropical *Pterombrus* with description of the diapausing larva (Hymenoptera: Tiphiidae). *Proc. Entomol. Soc. Wash.* 78(3): 361–368.

Kukla, G. J. 1972. The end of the present interglacial. *Quat. Res.* 2: 261–269.

Kurz, H. 1942. Florida dunes and scrub, vegetation and geology. *Fla. Geol. Surv. Geol. Bull.* 23: 1–154.

Kuster, J. E. 1979. Comparative structure of compound eyes of Cicindelidae and Carabidae (Coleoptera): Evolution of scotopy and photopy. *Quaest. Entomol.* 15(3): 297–334.

———. 1980. Fine structure of the compound eyes and interfacetal mechanoreceptors of *Cicindela tranquebarica* Herbst (Coleoptera: Cicindelidae). *Cell Tissue Res.* 206: 123–138.

Kuster, J. E., and W. G. Evans. 1980. Visual fields of the compound eyes of four species of Cicindelidae (Coleoptera). *Can. J. Zool.* 58(3): 326–336.

LaBonte, J. R., and P. J. Johnson. 1988. Millipede predation by *Omus dejeani* Reiche. *Cicindela* 20(3/4): 53–54.

Laessle, A. M. 1942. *The plant communities of the Welaka area, with special reference to correlations between soils and vegetational succession.* University of Florida, Biological Sciences Series 4, no. 1. 143 pp.

———. 1958. The origin and successional relationships of sandhill vegetation and sand-pine scrub. *Ecol. Monogr.* 28(4): 361–387.

———. 1967. Relationship of sand pine scrub to former shore lines. *Q. J. Fla. Acad. Sci.* 30(4): 268–286.

Laferté-Sénectère, F. T. 1841. Desc. de 10 Carabiques n. de Texas. *Rev. Zool.* 4: 37.

Landry, J. F., J. F. Rancourt, and P. Belanger. 1975. *Cicindela lepida* Dejean, un Carabidae nocturne. *Cordulia* 1(4): 101.

La Rivers, I. 1946. An annotated list of the Cicindelidae known to occur in Nevada (Coleoptera). *Pan-Pac. Entomol.* 22(4): 135–141.

Larochelle, A. 1974a. Observations on the biology of some tiger beetles [Cicindelidae]. *Entomol. News* 85(4): 99–101.

———. 1974b. North American amphibians and reptiles as predators of tiger beetles. *Cicindela* 6(4): 83–86.

———. 1976. Les cicindeles du Quebec. *Fabreries* 2(6): 72–79.

———. 1977a. Cicindelid papers published in *Entomological News* (1890–1974). *Cicindela* 9(4): 65–73.

———. 1977b. Cicindelidae caught at lights. *Cicindela* 9(3): 50–60.

———. 1977c. Notes on the food of tiger beetle larvae. *Cicindela* 9: 13–14.

———. 1978a. A bibliography of cicindelid papers published in *The Entomologists' Monthly Magazine* (1864–1972). *Cicindela* 10(1): 1–6.

———. 1978b. A bibliography of papers on Cicindelidae published in *The Entomologist* (1840–1973). *Cicindela* 10(2): 17–20.

———. 1978c. Further notes on birds as predators of tiger beetles. *Cicindela* 10(3): 37–41.

———. 1978d. Techniques for catching tiger beetles. *Cicindela* 10(2): 23–26.

———. 1978e. Catalogue des parasites et des phoretiques animaux des coleopteres Carabidae (les Cicindelini compris) du monde (Derniere partie). *Cordulia* 4(4): 69–74.

———. 1978f. Catalogue des parasites et phoretiques animaux des coleopteres Carabidae (les Cicindelini compris) du monde (Premiere partie). *Cordulia* 4(1): 1–7.

———. 1978g. Le cannibalisme chez les coleopteres Carabidae (les Cicindelini inclus). *Cordulia* 4(3): 89–92.

———. 1978h. Les plantes insectivores, predateurs de coleopteres Carabidae (les Cicindelini compris). *Cordulia* 4(2): 65.

———. 1978i. A bibliography of identification keys to European tiger beetles (1879–1974). *Cicindela* 10(3): 43–47.

———. 1978j. C. W. Leng's contributions to cicindelid entomology. *Cicindela* 10(3): 33–36.

———. 1979. Cicindelidae from the U.S.A. in the Canadian National Collection. *Cicindela* 11(1): 13–15.

———. 1980a. A preliminary list of papers dealing with the tiger beetles (Coleoptera: Carabidae: Cicindelini) of the world. Part 1. *Cordulia* 6(2–3): 25–60.

———. 1980b. A preliminary list of papers dealing with the tiger beetles (Coleoptera: Carabidae: Cicindelini) of the world. Part 2. *Cordulia* 6(4): 61–73.

———. 1980c. Bibliography of papers on Cicindelidae published in the journal *Psyche* (1874–1974). *Cicindela* 12(3): 40–42.

———. 1980d. Cicindelidae of the Maritime Provinces of Canada. *Cicindela* 12(3): 35–39.

———. 1985. A bibliography of papers on Cicindelidae published in the *Journal of the Kansas Entomological Society* (1928–1985). *Cicindela* 17(4): 67–68.

———. 1986a. A bibliography of papers on Cicindelidae published in the *Jour-

nal of the New York Entomological Society (1894–1979). *Cicindela* 18(3): 33–48.

———. 1986b. A bibliography of papers on Cicindelidae published in the *Pan-Pacific Entomologist* (1925–1979). *Cicindela* 18(1): 12–16.

———. 1986c. Cicindelidae from New England in the Museum of Comparative Zoology. *Cicindela* 18(4): 59–63.

———. 1986d. A concise bibliography on the geographical distribution of the Cicindelidae of North America north of Mexico. *Cicindela* 18(2): 17–32.

———. 1987a. A bibliography of papers on Cicindelidae published in the *Bulletin of the Brooklyn Entomological Society* (1878–1965). *Cicindela* 19(3–4): 51–59.

———. 1987b. A bibliography of papers on Cicindelidae published in the *Canadian Entomologist* (1869–1979). *Cicindela* 19(2): 21–36.

———. 1989a. A bibliography of papers published on Cicindelidae in the *Annual Reports and Proceedings of the Entomological Society of Ontario* (1871–1987). *Cicindela* 21(3/4): 41–47.

———. 1989b. A bibliography of papers published on Cicindelidae in the *Coleopterists Bulletin* (1947–1988). *Cicindela* 21(2): 17–26.

———. 1990. The food of carabid beetles (Coleoptera: Carabidae, including Cicindelinae). *Fabreries* 5 (Suppl.): 1–132.

Larson, D. J. 1986. The tiger beetle, *Cicindela limbata hyperborea* LeConte, in Goose Bay, Labrador (Coleoptera: Cicindelidae). *Coleopt. Bull.* 40(3): 249–250.

Larson, D. J., and D. W. Langor. 1982. The carabid beetles of insular Newfoundland (Coleoptera: Carabidae: Cicindelidae): 30 years after Lindroth. *Can. Entomol.* 114(7): 591–597.

Larson, P. R. 1981. The tiger beetles of North Dakota (Coleoptera: Cicindelidae). *Proc. N.D. Acad. Sci.* 35: 52.

La Rue, D. A. 1990. New California distribution records for two species of *Cicindela* Linnaeus (Coleoptera: Cicindelidae). *Cicindela* 22(4): 49–52.

———. 1994a. Additional distribution records for *Cicindela* (*Cicindelidia*) *nigrocoerulea nigrocoerulea* in California (Coleoptera: Cicindelidae). *Cicindela* 26(2): 25–26.

———. 1994b. Tiger beetles of the Algodones Sand Dunes, Imperial County, California (Coleoptera: Cicindelidae). *Cicindela* 26(1): 1–8.

Lavigne, R. J. 1972. Cicindelids as prey of robber flies (Diptera: Asilidae). *Cicindela* 4(1): 1–7.

———. 1977. Additional records of cicindelids as prey of robber flies (Diptera: Asilidae). *Cicindela* 9(2): 25–27.

Lawton, J. K. 1970. Notes on collecting tiger beetles in the southeastern United States. *Cicindela* 2(3): 1–7.

———. 1971. Collecting notes on the *Cicindela* of the Ozark Uplands, northwestern Louisiana, northwestern Texas, Oklahoma, and Kansas. *Cicindela* 3(4): 61–68.

———. 1972. Collecting notes on the tiger beetles in the southwestern and southcentral United States: spring trip 1970. *Cicindela* 4(2): 35–48.

LeConte, J. L. 1846. A descriptive catalogue of the geodephagous Coleoptera inhabiting the United States east of the Rocky Mountains. *Annals of the Lyceum of Natural History of New York.* 4: 173–474.

———. 1854. Note on the genus *Amblychila* Say. *Proc. Acad. Nat. Sci. Phila.* 7: 32–33.

———. 1857 (1856?). Revision of the Cicindelidae of the United States. *Trans. Am. Philos. Soc.* 11: 27–63.

———. 1868. New Coleoptera from Survey Kansas, N. Mexico. *Trans. Am. Entomol. Soc.* 2: 49–59.

———. 1875. Notes on the Cicindelidae of the United States. *Trans. Am. Entomol. Soc.* 5: 157–162.

———. 1856(7). Revision of the Cicindelidae of the United States. *Trans. Am. Philos. Soc.* 11: 27–63.

Lee, J. H., S. J. Johnson, and V. L. Wright. 1990. Quantitative survivorship analysis of the velvetbean caterpillar (Lepidoptera: Noctuidae) pupae in soybean fields in Louisiana. *Environ. Entomol.* 19(4): 978–986.

Leffler, S. R. 1973. Tiger beetles of Assateague Island National Seashore, Virginia. *Cicindela* 5(2): 35–37.

———. 1979a. A new subspecies of *Cicindela bellissima* from northwestern Washington (Coleoptera: Cicindelidae). *Coleopt. Bull.* 33(4): 465–472.

———. 1979b. Tiger beetles of the Pacific Northwest (Coleoptera: Cicindelidae). Dissertation Abstracts International, Section B, Sciences and Engineering 40(6): 2517.

———. 1980a. *Cicindela purpurea hatchi*, a replacement name for the preoccupied *C. p. mirabilis* Casey (Coleoptera: Cicindelidae). *Coleopt. Bull.* 34(1): 128.

———. 1980b. The larva of *Mantichora* Fabricius. *Cicindela* 12(1): 1–12.

———. 1981. Contributions to the knowledge of Cicindelidae by Louis-Jerome Reiche. *Cicindela* 13(3/4): 37–42.

———. 1985a. *Omus submetallicus* G. Horn: historical perspective, systematic position, type locality, and habitat. *Cicindela* 17(3): 37–50.

———. 1985b. The tiger beetle genus *Omus* Eschscholtz: larval characters and their implications. *Cicindela* 17(4): 53–66.

———. 1987. Synonymic notes on, and additions to, "Notes on cicindelid habitats in Oregon" by Maser and Beer (1984). *Cicindela* 19(1): 1–12.

Leffler, S. R., R. E. Nelson, and E. van den Berghe. 1986. Color variation and sex ratio in *Omus dejeani* Reiche. *Cicindela* 18(1): 7–11.

Leng, C. W. 1884. Cicindelidae of Staten Island. *Proc. Staten Island Assoc.* Jan.: 12.

———. 1902a. Notes on the Cicindelidae of Louisiana. *J. N.Y. Entomol. Soc.* 10(Sept.): 131–136.

———. 1902b. Notes on the Cicindelidae of the Pine Barrens of New Jersey. *J. N.Y. Entomol. Soc.* 10(Dec.): 236–240.

———. 1902c. Revision of Cicindelidae of boreal America. *Trans. Am. Entomol. Soc.* 28: 93–186.

———. 1910. Notes on Coleoptera collected in northern Georgia. *J. N.Y. Entomol. Soc.* 18(2): 71–82.

———. 1912. The geographical distribution of Cicindelidae in eastern North America. *J. N.Y. Entomol. Soc.* 20(1): 1–17.

———. 1915. List of the Carabidae of Florida. *Bull. Am. Mus. Nat. Hist.* 34: 555–601.

———. 1918. A new race of *Cicindela* with notes on other races and species. *J. N.Y. Entomol. Soc.* 26(3–4): 138–141.

———. 1920. *Catalogue of the Coleoptera of America north of Mexico.* Mount Vernon, N.Y.: John D. Sherman, Jr. 470 pp.

Leng, C. W., and W. Beutenmuller. 1894a. Preliminary handbook of the Coleoptera of Northeastern America. *J. N.Y. Entomol. Soc.* 2: 87–96.

———. 1894b. Additions and corrections to the list of Cicindelidae and Carabidae. *J. N.Y. Entomol. Soc.* 2(1): 43–48.

Leng, C. W., and W. T. Davis. 1924. List of the Coleoptera of Staten Island. *Proc. Staten Island Inst. Arts Sci.* 2: 1–82.

Leng, C. W., and A. J. Mutchler. 1914. A preliminary list of the Coleoptera of the West Indies as recorded to Jan. 1, 1914. *Bull. Am. Mus. Nat. Hist.* 33(30): 391–493.

———. 1916. Descriptive catalogue of West Indian Cicindelidae. *Bull. Am. Mus. Nat. Hist.* 35(36): 681–699.

———. 1927. *Catalogue of the Coleoptera north of Mexico. Supplement 1919–1924 (inclusive).* Mount Vernon, N.Y.: John D. Sherman, Jr. 78 pp.

———. 1933. *Catalogue of the Coleoptera north of Mexico. Second and third supplements 1925–1932 (inclusive).* Mount Vernon, N.Y.: John D. Sherman, Jr. 112 pp.

Leonard, J. G., and R. T. Bell. 1999. *Northeastern tiger beetles: a field guide to tiger beetles of New England and eastern Canada.* Boca Raton, Fla.: CRC Press. xii + 176 pp.

Lidicker, W. A. 1962. The nature of subspecies boundaries in desert rodents and its implications for subspecies taxonomy. *Syst. Zool.* 11(4): 160–171.

Liebeck, C. 1890. Cicindelidae of a season. *Entomol. News* 1(10): 158–160.

Lindroth, C. H. 1953. Some attempts toward experimental zoogeography. *Ecology* 34: 657–666.

———. 1963. The ground beetles (Carabidae, excl. Cicindelinae) of Canada and Alaska. Part 3. *Opusc. Entomol.* Suppl. 24: 201–408.

———. 1963. The problem of late land connections in the North Atlantic areas, pp. 73–85. *In:* A. Löve, ed., *North Atlantic biota and their history.* Oxford: Pergamon Press.

———. 1969. The theory of glacial refugia in Scandinavia. *Not. Entomol.* 49: 178–192.

———. 1974. On the elytral microsculpture of carabid beetles. *Entomol. Scand.* 5(3–4): 251–264.

Löding, H. P. 1945. Catalogue of the beetles of Alabama. *Geological Survey of Alabama Monograph* 2: 1–172.

Lowe, C. H., Jr. 1950. The systematic status of the *salamander Plethodon hardii*, with a discussion of biogeographical problems in Aneides. *Copeia* 1950: 92–99.

———. 1955. An evolutionary study of the island faunas in the Gulf of California, and Mexico, with a method for comparative analysis. *Evolution* 9: 339–344.

Lundelius, E. L. 1967. Late Pleistocene and Holocene faunal history of central Texas, pp. 287–319. *In:* P. S. Martin and H. E. Wright, Jr., eds., *Pleistocene extinctions, the search for a cause.* Proceedings of the VII Congress of the International Association for Quaternary Research, vol. 6. New Haven: Yale University Press.

MacArthur, R. H. 1962. Some generalized theorems of natural selection. *Proc. Natl. Acad. Sci. U.S.A.* 48: 1893–1897.

MacArthur, R. H., and E. Pianka. 1966. On optimal use of a patchy environment. *Am. Nat.* 100: 603–609.

MacArthur, R. H., and E. O. Wilson. 1963. An equilibrium theory of insular zoogeography. *Evolution* 17: 373–387.

———. 1967. *The theory of island biogeography.* Monographs in population biology 1. Princeton, N.J.: Princeton University Press.

MacArthur, R. H., and R. Levins. 1964. Competition, habitat selection and character displacement in a patchy environment. *Proc. Natl. Acad. Sci. U.S.A.* 51: 1207–1210.

MacArthur, R. H., H. Recher, and M. Cody. 1966. On the relation between habitat selection and species diversity. *Am. Nat.* 100: 319–327.

MacNamara, C. 1922. Popular and practical entomology. Tiger beetle larvae. *Can. Entomol.* 54:241–246.

MacNeil, F. S. 1950. Pleistocene shorelines in Florida and Georgia. U.S. Geological Survey Professional Paper 221–F: 91–107.

Magillivray, A. D. 1923. *External insect anatomy, a guide to the study of the insect anatomy and an introduction to systematic entomology.* Urbana, Ill.: Scarab Co. 388 pp.

Mandl, K. 1954. Aedeagus-Studien an Cicindeliden-Gattung (Col.). *Entomologische Arbeiten aus dem Museum G. Frey Tutzing Bei Muenchen* 5: 1–19.

———. 1955. Die Cicindeliden, Caraben und Calosomen der Afghanistanexpedition 1951 und 1952 J. Klapperichs. *Entomologische Arbeiten aus dem Museum G. Frey Tutzing Bei Muenchen* 6: 317–324.

———. 1956. Neun neue Cicindelidenformen aus tropischen Landern. *Entomologische Arbeiten aus dem Museum G. Frey Tutzing Bei Muenchen* 7: 378–397.

———. 1971. Restoration of the family status of the Cicindelidae (Coleoptera) (in German; English summary). *Beitraege zur Entomologie* 21(3/6): 507–508.

———. 1974. New Cicindelidae forms from South America (Coleoptera) [*Megacephala huedepohli, Cicindela trifasciata peruviana, Cicindela trifasciata microsoma*]. *Zeitschrift der Arbeitsgemeinschaft Oesterreichischer Entomologen* 26(1): 15–22.

Martin, J. O. 1932. *Amblycheila* in California. *Pan-Pac. Entomol.* 8: 111.

Martin, P. S., and B. E. Harrell. 1957. The Pleistocene history of temperate biotas in Mexico and eastern United States. *Ecology* 38(3): 468–480.

Maser, C. 1971. A simple method for preserving larval Cicindelidae. *Cicindela* 3: 79.

———. 1977a. Notes on *Cicindela oregona* in Oregon. *Cicindela* 9(4): 61–64.

———. 1977b. Notes on *Omus audouini* in Oregon. *Cicindela* 9(3): 47–49.

———. 1977c. Notes on *Omus dejeani* [Oregon coast]. *Cicindela* 9(2): 35–38.

Maser, C., and F. M. Beer. 1984. Notes on cicindelid habitats in Oregon. *Cicindela* 16(3–4): 39–60.

Maslin, T. P. 1952. Morphological criteria of phyletic relationships. *Syst. Zool.* 1: 49–70.

Mason, H. L. 1942. Distributional history and fossil record of *Ceanothus*. University of California Publ. Zoology. 50: 281–303.

Masteller, E. C., G. Kedzierski, and S. Spichiger. 1993. Terrestrial arthropods of Presque Isle State Park, Erie County, Pennsylvania, including Pseudoscorpiones, Opiliones, Isopoda, Diplopoda, Chilopoda and selected insects. *J. Pa. Acad. Sci.* 67(3): 127–131.

Mateu, J. 1975. A new *Amblychila* Say from the high tableland of Mexico (Coleoptera, Cicindelidae). *Anales de la Escuela Nacional de Ciencias Biologicas Mexico* 21(1/4): 145–153.

Mather, B. 1971. A preliminary survey of Mississippi Cicindelidae. *Cicindela* 3(2):21–25.

Matthew, N. D. 1906. Hypothetical outlines of the continents in Tertiary times. *Bull. Am. Mus. Nat. Hist.* 22: 359–361.

Maxwell, J. A., and M. B. Davis. 1972. Pollen evidence of Pleistocene and Holocene vegetation on the Allegheny Plateau, Maryland. *Quat. Res.* 2: 506–530.

Maxwell, J. C. 1968. Continental drift and a dynamic earth. *Am. Sci.* 1: 35–51.

May, M. L., D. L. Pearson, and T. M. Casey. 1986. Oxygen consumption of active and inactive adult tiger beetles. *Physiol. Entomol.* 11(2): 171–179.

Mayne, W. 1949. The foraminiferal genus *Choffatella* Schlumberger in the Lower Cretaceous of the Caribbean region (Venezuela, Cuba, Mexico, and Florida). *Eclogae Geologicae Helvetiae* 42: 529–547.

Mayr, E. 1947. Ecological factors in speciation. *Evolution* 1: 263–288.

———. 1974. *Populations, species, and evolution; an abridgement of Animal Species and Evolution*. 3d ed. Cambridge, Mass.: Harvard University Press, Belknap Press. 453 pp.

McCarley, W. H. 1954. Natural hybridization in the *Peromyscus leucopus* species group of mice. *Evolution* 8: 314–323.

McCluskey, R. 1993. Collecting Cicindelidae by mowing the lawn. *Cicindela* 25(1–2): 27–28.

McConker, E. H. 1954. A systematic study of the North American lizards of the genus *Ophisaurus*. *Am. Midl. Nat.* 51: 133–171.

McCrone, J. D. 1963. Taxonomic status and evolutionary history of the *Geolycosa pikei* complex in the southeastern United States (Aranaea: Lycosidae). *Am. Midl. Nat.* 70(1): 47–73.

McCrone, J. D., and H. W. Levi. 1964. North American widow spiders of the *Latrodectus curacaviensis* group (Araneae: Theridiidae). *Psyche* 71(1): 12–27.

Mengel, R. M. 1964. The probable history of species formation in some northern wood warblers. *Living Bird* 3: 9–43.

Meserve, F. G. 1936. The Cicindelidae of Nebraska (Coleoptera). *Entomol. News* 47(10): 270–275.

Meyerhoff, A. A., and H. A. Meyerhoff. 1972a. Continental drift, IV: the Caribbean "plate." *J. Geol.* 80: 34–60.

———. 1972b. The new global tectonics: major inconsistencies. *Am. Assoc. Pet. Geol. Bull.* 56(2): 269–336.

Miller, R. 1950. Ecological comparisons of plant communities of the xeric, pine type on sand ridges in central Florida Master's thesis, University of Florida, Gainesville.

Miskimen, G. W. 1961. Zoogeography of the coleopterous family Chauliognathidae. *Syst. Zool.* 10: 140–153.

Mitchell, J. D. 1903. Observations on the habits of two Cicindelidae. *Proc. Entomol. Soc. Wash.* 5: 108–110.

Montgomery, B. E., and R. W. Montgomery. 1931. Records of Indiana Coleoptera. I. Cicindelidae. *Proc. Indiana Acad. Sci.* 40: 357–359.

Mooi, R., P. F. Cannell, V. A. Funk, P. M. Mabee, and C. K. Starr. 1989. Historical perspectives, ecology, and tiger beetles: an alternative discussion. *Syst. Zool.* 38(2): 191–195.

Moore, B. P., and W. V. Brown. 1971. Benzaldehyde in the defensive secretion of a tiger beetle (Coleoptera: Carabidae). *J. Aust. Entomol. Soc.* 10(2): 142–143.

Moore, J. C. 1960. Squirrel geography of the Indian subregion. *Syst. Zool.* 9: 1–17.

Moore, R. 1906. Notes on the habits of *Cicindela*. *Entomol. News* 17: 338–343.

Morgan, A. V., and R. Freitag. 1981. The occurrence of *Cicindela limbalis* Klug (Coleoptera: Cicindelidae) in a late-glacial site at Brampton, Ontario. *Coleopt. Bull.* 36(1): 105–108.

Morgan, K. R. 1985. Body temperature regulation and terrestrial activity in the ectothermic beetle *Cicindela tranquebarica*. *Physiol. Zool.* 58(1): 29–37.

Mossakowski, D. 1979. Reflection measurements used in the analysis of structural colours of beetles. *J. Microsc.* 116(3): 351–364.

———. 1984. Cuticle structure and colour in *Cicindela*. *Int. Congr. Entomol. Proc.* 17: 80.

Muller, H. J. 1942. Isolating mechanisms, evolution, and temperature. *Biol. Symp.* 6: 71–125.

Mulvania, M. 1931. Ecological survey of a Florida scrub. *Ecology* 12: 528–540.

Munroe, E. 1965. Zoogeography of insects and allied groups. *Annu. Rev. Entomol.* 10: 325–344.

Murray, G. E., and A. E. Weidie, Jr. 1967. Regional geologic summary of Yucatan Peninsula, pp. 5–50. *In* A. E. Weidie, Jr., ed., *Field trip to Peninsula of Yucatan, Guidebook*. New Orleans: New Orleans Geological Society.

Murray, R. R. 1979. The *Cicindela* fauna of Mexico: range extensions, additions, and ecological notes (Coleoptera: Cicindelidae). *Coleopt. Bull.* 33(1): 49–56.

———. 1980. Systematics of *Cicindela rufiventris* Dejean, *Cicindela sedecimpunctata* Klug and *Cicindela flohri* Bates (Coleoptera: Cicindelidae). Ph.D. diss., Texas A&M University, College Station. 287 pp.

———. 1983. Redescription of *Cicindela speculans* Bates and its relationship to other Neotropical *Cicindela* (Coleoptera: Cicindelidae). *Entomol. News* 94(3): 81–85.

Mury Meyer, E. J. 1981. The capture efficiency of flickers preying on larval tiger beetles. *Auk* 98(1): 189–191.

———. 1983. An analysis of survivorship and foraging methods in the larvae of three sympatric species of tiger beetles Coleoptera: Cicindelidae) occurring in Central Pennsylvania. Ph.D. diss., Pennsylvania State University, University Park. 127 p.

———. 1987. Asymmetric resource use in two syntopic species of larval tiger beetles (Cicindelidae). *Oikos* 50(2): 167–175.

Mutchler, A. J. 1924. A new species of Cicindelidae from Cuba. *Am. Mus. Novit.* 106: 1–3.

Mutchler, A. J., and C. W. Leng. 1916. Descriptive catalogue of West Indian Cicindelinae. *Bull. Am. Mus. Nat. Hist.* 35: 681–699.

Myers, R. L., and J. J. Ewel, eds. 1990. *Ecosystems of Florida*. Orlando: University of Central Florida Press. xviii + 765 p.

Nagano, C. D. 1980a. Black widow spider feeds on tiger beetle. *Cicindela* 12(2):28.

———. 1980b. Population status of the tiger beetles of the genus *Cicindela* (Coleoptera: Cicindelidae) inhabiting the marine shoreline of southern California. *Atala* 8(2): 33–42.

———. 1985. Distributional notes on the tiger beetles of the California Channel Islands (Coleoptera: Cicindelidae), pp. 105–112. *In:* A. S. Menke, and D. R. Miller, eds., *Entomology of the California Channel Islands: Proceedings of the first Symposium*. Santa Barbara, Calif.: Santa Barbara Museum of Natural History. 178 pp.

Nagano, C. D., S. E. Miller, and A. V. Morgan. 1982. Fossil tiger beetles (Coleoptera: Cicindelidae): review and new Quaternary records. *Psyche* 89(3–4): 339–346.

Neill, W. T. 1957. Historical biogeography of present-day Florida. *Fla. State Mus. Bull.* 2: 175–220.

Nelson, R. E., and J. R. LaBonte. 1989. Rediscovery of *Cicindela ancociscoensis* T. W. Harris and first records for *C. scutellaris lecontei* Haldeman in Maine. *Cicindela* 21(3/4): 49–54.

Newman, E. 1838. Description of *Cicindela ventralis*, p. 414. *In:* Art. XLIII. Communications on the natural history of North America, by E. Doubleday. *Entomol. Mag.* 5: 409–417.

Nicolay, A. S., and H. B. Weiss. 1932. Synopsis of the Cicindelidae. I. General introduction, bibliography and purpurea group. *J. N.Y. Entomol. Soc.* 40(3): 341–355.

Niemala, J., E. Ranta, D. L. Pearson, and S. A. Juliano. 1993. World-wide tiger beetle mandible length ratios: was something left unmentioned? *Ann. Zool. Fenn.* 30(1): 85–88.

Noonan, G. R. 1973. The Anisodactylines (Insecta: Coleoptera: Carabidae Harpalini): classification, evolution, and zoogeography. *Quaest. Entomol.* 9: 266–480.

———. 1979. The science of biogeography with relation to Carabids, pp. 1–12. *In:* T. L. Erwin, G. E. Ball, D. R. Whitehead, and A. L. Halpern, eds., *Carabid beetles: their evolution, natural history, and classification*. The Hague: Junk. 635 pp.

Noonan, G. R., G. E. Ball, and N. E. Stork, eds. 1992. *The biogeography of ground beetles of mountains and islands*. Andover, Hampshire: Intercept. 256 pp.

Nordin, P. D. 1985. Interspecific mating of two alpine California cicindelids. *Cicindela* 17(1): 13–15.

Norris, R. A. 1958. Comparative biosystematics and life history of the nuthatches *Sitta pygmaea* and *S. pusilla*. *Univ. Calif. Publ. Zool.* 56: 119–300.

Nunenmacher, F. W. 1940. Studies on the species of *Omus*, No. 1. *Pan-Pac. Entomol.* 16: 143–144.

Olsen, S. J. 1965. Vertebrate fossil localities in Florida. Tallahassee: Florida Geological Survey Special Publication 12.

Olson, A. L., T. H. Hubbell, and H. T. Howden. 1954. The burrowing beetles of genus *Mycotrupes*. *Misc. Publ. Mus. Zool. Univ. Mich.* 84: 1–59.

Osgood, W. H. 1909. Revision of the mice of the American genus *Peromyscus*. *North American Fauna.* 28: 1–285.

Paarman, W. 1977. Propagation rhythm of subtropical and tropical Carabidae and its control by exogenous factors. *Adv. Invertebr. Reprod.* 1: 49–60.

Pajni, H. R., and S. S. Bedi. 1978a. The female genitalia of family Cicindelidae and its taxonomic significance. *Res. Bull. Panjab Univ. Sci.* 29(1–4): 23–27.

———. 1978b. Structure of the male genitalia of the family Cicindelidae (Coleoptera). *Res. Bull. Panjab Univ. Sci.* 29(1–4): 39–44.

Palmer, M. K. 1976. Notes on the biology of *Pterombus piceus* Krombein (Hymenoptera: Tiphiidae). *Proc. Entomol. Soc. Wash.* 78(3): 369–375.

———. 1978. Growth rates and survivorship of tiger beetle larvae. *Cicindela* 10(4): 49–66.

———. 1979. Rearing tiger beetles in the laboratory. *Cicindela* 11(1): 1–11.

———. 1981. Notes on the biology and behavior of *Odontochila mexicana*. *Cicindela* 13(3–4): 29–36.

———. 1982. Biology and behavior of two species of *Anthrax* (Diptera: Bombyliidae), parasitoids of the larvae of tiger beetles (Coleoptera: Cicindelidae). *Ann. Entomol. Soc. Am.* 75(1): 61–70.

———. 1983. *Pterombrus piceus* (avispa escara bajo, tiphiid wasp, beetle wasp), p. 766. *In:* D. H. Janzen, ed., *Costa Rican natural history.* Chicago: University of Chicago Press.

Palmer, M. K., and M. A. Gorrick. 1979. Influence of food on development in tiger beetle larvae. *Cicindela* 11(2): 17–25.

Palmer, R. K. 1945. Outline of the geology of Cuba. *J. Geol.* 53: 1–34.

Papp, H. 1952. Morphologische und phylogenetische untersuchungen an *Cicindela*—arten. Unter besonderer Berucksichtigung der Ableitung der nearktischen formen. *Oesterr. Zool. Z.* 3: 494–533.

Park, O. 1949. Application of the converse Bergmann principle to the Carabid beetle, *Dicaelus purpuratus*. *Physiol. Zool.* 22(4): 259–372.

Parker, K. C. 1955. Sympatry, allopatry, and the subspecies in birds. *Syst. Zool.* 4 (1): 35–40.

Patil, B. V., M. C. Devaiah, and T. S. Thontadarya. 1982. Studies on the attraction of predatory insects to mercury bulb light-trap. *Indian J. Ecol.* 9(1): 108–112.

Patterson, J. T., and A. B. Griffen. 1944. A genetic mechanism underlying species in isolation. University of Texas Publication 4445: 212–223.

Pearson, D. L. 1980. Patterns of limiting similarity in tropical forest tiger beetles (Coleoptera: Cicindelidae). *Biotropica* 12(3): 195–204.

———. 1985. The function of multiple anti-predator mechanisms in adult tiger beetles (Coleoptera: Cicindelidae). *Ecol. Entomol.* 10(1): 65–72.

———. 1988. Biology of tiger beetles. *Annu. Rev. Entomol.* 33: 123–147.

———. 1992. Tiger beetles as indicators for biodiversity patterns in Amazonia. *Res. Explor.* 8(1): 116–117.

Pearson, D. L., and J. J. Anderson. 1985. Perching heights and nocturnal communal roosts of some tiger beetles (Coleoptera: Cicindelidae) in southeastern Peru. *Biotropica* 17(2): 126–129.

Pearson, D. L., M. S. Blum, T. H. Jones, H. M. Fales, E. Gonda, and B. R. Witte. 1988. Historical perspective and the interpretation of ecological patterns: defensive compounds of tiger beetles (Coleoptera: Cicindelidae). *Am. Nat.* 132(3): 404–416.

Pearson, D. L., and F. Cassola. 1992. World-wide species richness patterns of tiger beetles (Coleoptera: Cicindelidae): indicator taxon for biodiversity and conservation studies. *Conserv. Biol.* 6(3): 376–391.

Pearson, D. L., and S. A. Juliano. 1991. Mandible length ratios as a mechanism for co-occurrence: evidence from a world-wide comparison of tiger beetle assemblages (Cicindelidae). *Oikos* 61(2): 223–233.

———, eds. 1993. *Evidence for the influence of historical processes in co-occurrence and diversity of tiger beetle species.* Chicago: University of Chicago Press. 414 pp.

Pearson, D. L., and C. B. Knisley. 1985. Evidence for food as a limiting resource in the life cycle of tiger beetles (Coleoptera: Cicindelidae). *Oikos* 45(2): 161–168.

Pearson, D. L., and R. C. Lederhouse. 1987. Thermal ecology and the structure of an assemblage of adult tiger beetle species (Cicindelidae). *Oikos* 50(2): 247–255.

Pearson, D. L., and E. J. Mury. 1978. Possible character divergence of mandible size and gape in sympatric tiger beetles (Coleoptera: Cicindelidae). *J. N.Y. Entomol. Soc.* 85(4): 194–195.

———. 1979. Character divergence and convergence among tiger beetles (Coleoptera: Cicindelidae). *Ecology* 60(3): 557–566.

Pearson, D. L., and S. L. Stemberger. 1980. Competition, body size and the relative energy balance of adult tiger beetles (Coleoptera: Cicindelidae). *Am. Midl. Nat.* 104(2): 373–377.

Pearson, D. L., and A. P. Vogler. 2001. *Tiger beetles: the evolution, ecology, and diversity of the Cicindelids.* Ithaca, N.Y.: Comstock Publishing Associates, Cornell University Press. xiii + 333 pp.

Peck, S. B., and M. C. Thomas. 1998. *A distributional checklist of the beetles (Coleoptera) of Florida.* Vol. 16, *Arthropods of Florida and neighboring land areas.* Florida Department of Agriculture and Consumer Services, Division of Plant Industry. Entomology Contribution 862. viii + 180 pp.

Peña, L. E., and G. Barria. 1973. Revision de la familia Cicindelidae en Chile. *Rev. Chil. Entomol.* 7: 183–191.

Peyton, E. L., J. F. Reinert, and N. Peterson. 1964. The occurrence of *Deinocerites pseudes* Dyar and Knab in the United States, with additional notes on the biology of *Deinocerites* species of Texas. *Mosq. News* 24(4): 449–458.

Phillips, J. D., and D. Forsyth. 1972. Plate tectonic, paleomagnetism, and the opening of the Atlantic. *Geol. Soc. Am. Bull.* 83: 1579–1600.

Pianka, E. R. 1966. Latitudinal gradients in species diversity: a review of concepts. *Am. Nat.* 100: 33–46.

Pirkle, E. C., W. H. Yoho, and C. W. Hendry, Jr. 1970. Ancient sea level stands in Florida. Florida Department of Natural Resources Bureau of Geology Bulletin 52. vii + 61 pp.

Pitelka, F. A. 1951. Speciation and ecologic distribution in American jays of the genus *Aphelocoma*. *Univ. Calif. Publ. Zool.* 50: 195–464.

Platnick, N. I. 1976a. Drifting spiders or continents? Vicariance biogeography of the spider subfamily Laroniinae (Araneae: Gnaphosidae). *Syst. Zool.* 25: 101–109.

———. 1976b. Concepts of dispersal in historical biogeography. *Syst. Zool.* 25: 294–295.

Platnick, N. I., and G. Nelson. 1978. A method of analysis for historical biogeography. *Syst. Zool.* 27: 1–16.

Poerinck, W. H. J. 1953. Caribbean tiger beetles of the genus *Cicindela*. *Studies on the fauna of Curaçao and other Caribbean islands* 4(19): 120–143.

Pomeroy, A. W. 1932. African beetles of the family Carabidae. *Trans. Entomol. Soc. Lond.* 80: 77–103.

Porter, S. C., and G. H. Denton. 1967. Chronology of neoglaciation in the North American Cordillera. *Am. J. Sci.* 265: 177–210.

Potts, R. W. L. 1943. Habits of *Amblycheila cylindriformis* Say. *Pan-Pac. Entomol.* 19(3): 85.

Potzger, J. E. and B. C. Thorp. 1947. Pollen profile from a Texas bog. *Ecology* 28: 274–280.

Pratt, R. Y. 1939. The mandibles of *Omus dejeani* Reiche as secondary sexual organs. *Pan-Pac. Entomol.* 15(2): 95–96.

Preston, F. W. 1962. The canonical distribution of commonness and rarity. *Ecology* 43: 185–215, 410–432.

Puri, H. S. 1953. Contribution to the study of the Miocene of the Florida panhandle. *Fla. Geol. Surv. Bull.* 36: 1–345.

Puri, H. S., and R. O. Vernon. 1964. Summary of the geology of Florida, and a guidebook to the classic exposures. Florida State Board of Conservation, Division of Geological Survey, Special Publication 5, revised.

Putchkov, A. V. 1990. Larvae of the cicindelid beetle subgenera *Lophyridia*, *Eugrapha*, *Cylindera* (Coleoptera, Carabidae) of the south-western European part of the USSR. *Vestn. Zool.* 4: 12–18.

———. 1993. Larvae of tiger beetles of the subgenus *Cicindela* s. str. (Coleoptera, Carabidae, Cicindelinae) from the Russian plain and the Caucasus. *Zool. Zh.* 72(7): 52–62.

———. 1994a. Larvae of the tiger beetles of the subgenus *Cicindela* s. str. (Coleoptera, Carabidae, Cicindelinae) from the Russian plains and the Caucasus. *Entomol. Rev.* 73(8): 141–151

———. 1994b. State-of-the-art and world perspectives of studies on tiger beetle larvae (Coleoptera, Carabidae, Cicindelinae). *Ser. Entomol.* 51: 51–54.

———. 1995. A description of the larval stages of *Megacephala (Grammognatha) euphratica armeniaca* Castelnau (Coleoptera: Cicindelidae). *Zeitschrift der Arbeitgemeinschaft Oesterreichischer Entomologen* 47(1–2): 36–40.

Putchkov, A. V., and E. Arndt. 1994. Preliminary list and key of known tiger beetle larvae (Coleoptera, Cicindelidae) of the world. *Mitteilungen der Schweizerischen Entomologischen Gesellschaft* 67(3–4): 411–420.

Putchkov, A. V., and F. Cassola. 1991. The larvae of tiger beetles from central Asia (Coleoptera, Cicindelidae): Studies on tiger beetles, LXXII. *Bollettino del Museo Civico di Storia Naturale di Verona* 18: 11–43.

Pyle, T. E. 1968. Late Tertiary history of the Gulf of Mexico based on core from Sigsbee Knolls. *Am. Assoc. Pet. Geol. Bull.* 53: 2501–2505.

Radovanovic, M. 1959. Zum problem der speziation bei Inseleidechsen. *Zoologische Jahrbuecher Systematik* 86: 395–436.

Raisz, E. 1964. *Atlas of Florida*. Gainesville: University of Florida Press. 52 pp.

Rand, A. L. 1948. Glaciation, a factor in speciation. *Evolution* 2(4): 314–321.

Ray, C. 1960. The application of Bergmann's and Allen's rules to the poikilotherms. *J. Morphol.* 106: 85–108.

Rees, N. E. 1982. Enemies of *Rhinocyllus conicus* in southwestern Montana. *Environ. Entomol.* 11(1): 157–158.

Reichardt, H. 1977. A synopsis of the genera of Neotropical Carabidae (Insecta: Coleoptera). *Quaest. Entomol.* 13: 346–493.

Reiche, L. J. 1838. Note sur le genre *Omus* d'Eschscholtz et description de deux nouvelle espèces. *Ann. Soc. Entomol. Fr.* 7: 297–302.

Rensch, B. 1938. Some problems of geographic variation and species formation. *Proc. Linn. Soc. Lond.* 150: 275–285.

Richards, H. G. 1963. Stratigraphy of earliest Mesozoic sediments in southeastern Mexico and western Guatemala. *Am. Assoc. Pet. Geol. Bull.* 47(11): 1861–1870.

Richards, H. G., and S. Judson. 1965. The Atlantic coastal plain and the Appalachian highlands in the Quaternary, pp. 129–136. *In:* H. E. Wright, Jr. and D. G. Frey, eds., *The quaternary of the United States, a review volume for the VII Congress of the International Association for Quaternary Research*. Princeton, N.J.: Princeton University Press.

Richman, D. B. 1967. Tiger beetle collecting in the Colorado desert. *Newsletter of the Association of Minnesota Entomologists* 1: 88–90.

Richmond, E. A. 1962. The fauna and flora of Horn Island, Mississippi. *Gulf Res. Rep.* 1(2): 1–106.

———. 1968. A supplement to the fauna and flora of Horn Island, Mississippi. *Gulf Res. Rep.* 2(3): 213–254.

Rivalier, E. 1948. Les cicindeles du genre *Lophyra* (Motsch.). *Rev. Fr. Entomol.* 15(2): 4–73.

———. 1950a. Retablissement de *Cicindela maroccana* Fab. dans su qualite d'espece. *Rev. Fr. Entomol.* 17(2): 93–96.

———. 1950b. Demembrement du genre *Cicindela* L. (Travail preliminaire limite a la fauna palearctique.) *Rev. Fr. Entomol.* 17(4): 217–244.

———. 1953a. Note sur une sous-espece meconnue de *Lophyridia angulata* Fab. *Rev. Fr. Entomol.* 20(1): 81–84.

———. 1953b. Les trois grandes sous-especes de *Lophyridia lunulata* Fab. *Rev. Fr. Entomol.* 20(3): 195–201.

———. 1954. Demembrement du genre *Cicindela* Linn. II. Faune américaine. *Rev. Fr. Entomol.* 21(4): 249–268.

———. 1955. Les *Brasiella* du groupe de *argentata* F. *Rev. Fr. Entomol.* 22: 77–100, 1 pl.

———. 1957. Demembrement du genre *Cicindela* L. III. Faune Africano-Malgache. *Rev. Fr. Entomol.* 24: 312–342.

———. 1961. Demembrement du genre *Cicindela* L. (Suite) (1). IV. Faune Indomalaise. *Rev. Fr. Entomol.* 28(3): 121–149.

———. 1963. Demembrement du genre *Cicindela* L. (fin.) V. Faune Australienne. (Et liste recapitulative des genres et sous-genres proposes pour la faune mondiale.) *Rev. Fr. Entomol.* 30(1): 30–48.

———. 1969. Demembrement du genre *Odontochila* et revision des principales especes. *Ann. Soc. Entomol. Fr.* 5(1): 195–237.

———. 1971. Remarque sur la tribu des Cicindelini et sa sub-division en soustribus. *Nouv. Rev. Entomol.* 1(2): 135–143.

———. 1972. Especies nouvelles du genre *Leptognatha* Riv. et du genre nouvelle *Orthocindela* de Nouvelle-Guinee. *Ann. Soc. Entomol. Fr.* 8(2): 299–307.

———. 1983. Chasse et preparation des cicindeles. *Bull. Soc. Sci. Nat.* 39: 7–8.

———. 1992. Dismemberment of the genus *Cicindela* Linnaeus. 2. American fauna (translation of Rivalier 1954). *Cicindela* 24(1–2): 7–42.

Rivers, J. J. 1893. The species of *Amblychila*. *Zoe Biol. J.* 4: 218–223.

Robinson, J. H., 1948. Description of a new tiger beetle from Texas. *Ann. Entomol. Soc. Am.* 41: 27.

Rogers, J. S. 1933. The ecological distribution of the craneflies of northern Florida. *Ecol. Monogr.* 3(1): 1–74.

———. 1955. Subspecies and clines: summary. *Syst. Zool.* 3: 126–133.

Roman, S. J. 1988. Collecting *Cicindela scabrosa* Schaupp, with notes on its habitat. *Cicindela* 20(2): 31–34.

Rosen, D. E. 1975. A vicariance model of Caribbean biogeography. *Syst. Zool.* 24: 431–464.

———. 1978. Vicariant patterns and historical explanation in biogeography. *Syst. Zool.* 27: 159–188.

Ross, E. S. 1956. What is species describing? *Syst. Zool.* 5: 191–192.

Ross, H. H. 1962. *A synthesis of evolutionary theory.* Englewood Cliffs, N.J.: Prentice-Hall. 387 pp.

———. 1963. The dunesland heritage of Illinois. *Ill. Nat. Hist. Surv. Circ.* 49: 1–28.

———. 1965a. Geological history of insects, pp. 418–438. *In:* H. H. Ross, *A textbook of entomology.* 3d ed. New York: John Wiley and Sons.

———. 1965b. Pleistocene events and insects, pp. 583–596. *In:* H. E. Wright, Jr. and D. G. Frey, eds., *The quaternary of the United States, a review volume for the VII Congress of the International Association for Quaternary Research.* Princeton, N.J.: Princeton University Press.

———. 1967. The evolution and past dispersal of the Trichoptera. *Annu. Rev. Entomol.* 12: 169–206.

———. 1974. *Biological systematics.* Reading, Mass.: Addison-Wesley. 345 pp.

Rotger, B. 1944. A new species of *Cicindela* and two new records of Coleoptera. *Pan-Pac. Entomol.* 20(2): 76–77.

———. 1972. A new race of *Cicindela willistoni* LeC. from New Mexico. *Cicindela* 4: 25–27.

———. 1974. A new subspecies of *Cicindela lengi* W. Horn. *Cicindela* 6: 9–11.

Rumpp, N. L. 1956. Tiger beetles of the genus *Cicindela* in southwestern Nevada and Death Valley, California, and description of two new subspecies (Coleoptera-Cicindelidae). *Bull. South. Calif. Acad. Sci.* 55: 144–154.

———. 1957. Notes on the *Cicindela praetextata-californica* tiger beetle complex. Description of a new subspecies from Death Valley, California (Coleoptera: Cicindelidae). *Bull. South. Calif. Acad. Sci.* 56: 144–154.

———. 1961. Three new tiger beetles of the genus *Cicindela* from south western United States. *Bull. South. Calif. Acad. Sci.* 60: 165–187.

———. 1967. A new species of *Cicindela* from Idaho (Coleoptera: Cicindelidae). *Proc. Calif. Acad. Sci.* 35(7): 129–140.

———. 1977. Tiger beetles of the genus *Cicindela* in the Sulphur Springs Valley, Arizona, with descriptions of three new subspecies (Cicindelidae: Coleoptera). *Proc. Calif. Acad. Sci.* 41(4): 169–181.

———. 1979. A note on *Cicindela longilabris* Say (Coleoptera: Cicindelidae). *Entomol. News* 90(1): 55.

———. 1986. Two new tiger beetles of the genus *Cicindela* from western United States (Cicindelidae: Coleoptera). *Bull. South. Calif. Acad. Sci.* 85(3): 139–151.

Russell, R. J., and R. D. Russell. 1939. Mississippi delta sedimentation in recent marine sediments, part 3. *Am. Assoc. Pet. Geol. Bull.* 23: 154–155.

Rutten, L. M. R. 1934. Geology of the Isla de Pinos, Cuba. *Proceedings of the Koninkliske Nederlandse Akademie van Wetenschappen* 37: 401–406.

———. 1940. On the age of serpentines in Cuba. *Proceedings of the Koninklijke Nederlandse Akademie van Wetenschappen* 43: 542–547.

Ryan, R. M. 1963. The biotic provinces of Central America. *Acta Zool. Mex.* 6: 1–54.

Sabrosky, C. W. 1940. Entomological usage of subspecific names. *Entomol. News* 51: 159–164.

Sailer, R. I. 1961. Utilitarian aspects of supergeneric names. *Syst. Zool.* 10: 154–156.

Say, T. 1817. Descriptions of several new species of North American insects. *J. Acad. Nat. Sci. Phila.* 1(2): 20.

———. 1818. A monograph of the North American insects of the genus *Cicindela*. *Trans. Am. Philos. Soc.*, New Series 1: 401–426.

Schaupp, F. G. 1878a. The Cicindelidae of the neighborhood of New York. *Bull. Brooklyn Entomol. Soc.* 1(4): 28.

———. 1878b. On the Cicindelidae of the United States (*Amblycheila*, *Omus*, and *Tetracha*). *Bull. Brooklyn Entomol. Soc.* 1: 11–14.

———. 1879a. Larvae of Cicindelidae. *Bull. Brooklyn Entomol. Soc.* 2:23–24.

———. 1879b. List of the described Coleopterous larvae of the United States with some remarks on their classification. *Bull. Brooklyn Entomol. Soc.* 2: 1–3, 21–22, 29–30.

———. 1883. Synoptic tables of Coleoptera. Cicindelidae. *Bull. Brooklyn Entomol. Soc.* 6(2): 73–108, 121–127.

———. 1884. Remarks and descriptions of new species. *Bull. Brooklyn Entomol. Soc.* 6: 121–124.

———. 1894. Bibliography. *Bull. Brooklyn Entomol. Soc.* 6: 85–108.

Schilder, F. A. 1953. Studien zur evolution von *Cicindela*. *Wissenschaftliche Zeitschrift Martin-Luther-Universitaet Halle-Wittenberg Mathematisch-Naturwissenschaftliche Reihe* 32: 539–576.

Schincariol, L. A., and R. Freitag. 1986. Copulatory locus, structure and function of the flagellum of *Cicindela tranquebarica* Herbst (Coleoptera: Cicindelidae). *Int. J. Invertebr. Reprod. Dev.* 9(3): 333–338.

———. 1991. Biological character analysis, classification, and history of the North American *Cicindela splendida* Hentz group taxa (Coleoptera: Cicindelidae). *Can. Entomol.* 123(6): 1327–1353.

Schlee, D. 1969. Hennig's principles of phylogenetic systematics, an intuitive, statistico-phenetic taxonomy. *Syst. Zool.* 18(1): 127–134.

Schmidt, J. O., and C. E. Mickel. 1979. A new *Dasymutilla* from Florida (Hymenoptera: Mutillidae). *Proc. Entomol. Soc. Wash.* 81(4): 576–579.

Schmidt, K. P. 1945. Evolution, succession, and dispersal. *Am. Midl. Nat.* 33: 788–790.

———. 1950. The concept of geographic range with illustration from amphibians and reptiles. *Tex. J. Sci.* 2: 326–334.

———. 1954. Faunal realms, regions, and provinces. *Q. Rev. Biol.* 29: 322–331.

Schneider, P. 1974. The start and flight of *Cicindela*. *Naturwissenschaften* 61(2): 82–83.

Schneider, P., and B. Kramer. 1974. Flight control in the tiger beetle (*Cicindela*) and the cockchafer (*Melolontha*). *J. Comp. Physiol.* 91(4): 377–386.

Schuchert, C. 1929. Geological history of the Antillean region. *Bull. Geol. Soc. Am.* 40: 337–360.

———. 1968. *Historical geology of the Antillean-Caribbean region, or the lands bordering the Gulf of Mexico and the Caribbean Sea*. New York: Hufner. 812 pp.

Schuler, L. 1960. Les spermatheques dan la tribu des Bembidiini Jeannel. *Rev. Fr. Entomol.* 27: 24–48.

———. 1963. La spermatheque chez les Harpalidae et les Pterostichitae de France. *Rev. Fr. Entomol.* 30(2): 81–103.

Schultz, C. B. 1972. Holocene interglacial migrations of mammals and other vertebrates. *Quat. Res.* 2: 337–340.

Schultz, T. D. 1981. Tiger beetles scavenging on dead vertebrates. *Cicindela* 13(3–4): 48.

———. 1982. Interspecific copulation of *Cicindela scutellaris* and *Cicindela formosa*. *Cicindela* 14(1–4): 41–44.

———. 1983. Opportunistic foraging of western kingbirds on aggregations of tiger beetles. *Auk* 100(2): 496–497.

———. 1984. The ultrastructure, physiology, and ecology of epicuticular interference reflectors in tiger beetle (*Cicindela*). Dissertation Abstracts International, Section B, Sciences and Engineering 45(3): 797.

———. 1986. Role of structural colors in predator avoidance by tiger beetles of the genus *Cicindela* (Coleoptera: Cicindelidae). *Bull. Entomol. Soc. Am.* 32(3): 142–146.

———. 1988. Destructive effects of off-road vehicles on tiger beetle habitat in central Arizona. *Cicindela* 20(2):25–29.

———. 1989. Habitat preferences and seasonal abundances of eight sympatric species of tiger beetle, genus *Cicindela* (Coleoptera: Cicindelidae), in Bastrop State Park, Texas. *Southwest. Nat.* 34(4): 468–477.

———. 1991. Tiger hunt. *Nat. Hist.* 100: 38–45.

———. 1994. Predation by larvae soldier beetles (Coleoptera:Cantharidae) on the eggs and larvae of *Pseudoxycheila tarsalis* (Coleoptera: Cicindelidae). *Entomol. News* 105(1): 14–16.

Schultz, T. D., and G. D. Bernard. 1989. Pointillistic mixing of interference colours in cryptic tiger beetles. *Nature* 337(6202): 72–73.

Schultz, T. D., and N. F. Hadley. 1987a. Microhabitat segregation and physiological differences in co-occurring tiger beetle species, *Cicindela oregona* and *Cicindela tranquebarica*. *Oecologia* 73(3): 363–370.

———. 1987b. Structural colors of tiger beetles and their role in heat transfer through the integument. *Physiol. Zool.* 60(6): 737–740.

Schultz, T. D., M. C. Quinian, and N. F. Hadley. 1992. Preferred body temperature, metabolic physiology, and water balance of adult *Cicindela longilabris:* a comparison of populations from boreal habitats and climatic refugia. *Physiol. Zool.* 65(1): 226–242.

Schultz, T. D., and M. A. Rankin. 1985a. Developmental changes in the interference reflectors and colorations of tiger beetles (*Cicindela*). *J. Exp. Biol.* 117: 111–117.

———. 1985b. The ultrastructure of the epicuticular interference reflectors of tiger beetles (*Cicindela*). *J. Exp. Biol.* 117: 87–110.

Schwarz, E. A. 1878. The Coleoptera of Florida. *Proc. Am. Philos. Soc.* 17: 353–469.

———. 1888. The insect fauna of semitropical Florida with special regard to the Coleoptera. *Entomol. Am.* 4(9): 165–177.

———. 1889. On a collection of Coleoptera from St. Augustine, Florida. *Proc. Entomol. Soc. Wash.* 1(3): 169–171.

Scudder, G. G. 1961. The comparative morphology of the insect ovipositor. *Trans. R. Entomol. Soc. Lond.* 113: 25–40.

Sears, P. B. 1942. Post glacial migration of five forest genera. *Am. J. Bot.* 29: 684–691.

Sharp, D., and F. N. Muir. 1912. The comparative anatomy of the male genital tube in Coleoptera. *Trans. Entomol. Soc. Lond.* 3: 477–642.

Shaw, P. B., J. C. Owens, E. W. Huddleston, and D. B. Richman. 1987. Role of arthropod predators in mortality of early instars of the range caterpillar, *Hemileuca oliviae* (Lepidoptera: Saturniidae). *Environ. Entomol.* 16(3): 814–820.

Shelford, V. E. 1907. Preliminary note on the distribution of the tiger beetles and its relation to plant succession. *Biol. Bull.* 14: 9–24.

———. 1908. Life-histories and larval habits of the tiger beetles (Cicindelidae). *J. Linn. Soc. Lond. Zool.* 30: 157–184.

———. 1911. Physiological animal geography. *J. Morphol.* 22: 551–618.

———. 1912. Ecological succession IV. Vegetation and the control of animal communities. *Biol. Bull.* 23: 59–99.

———. 1913a. Noteworthy variation in the elytral tracheation of *Cicindela*. *Entomol. News* 24: 124–125.

———. 1913b. The life-history of a bee-fly (*Spogostylum anale* Say) parasite of a tiger beetle (*Cicindela scutellaris* Say var. *lecontei* Hald.). *Ann. Entomol. Soc. Am.* 6: 213–225.

———. 1917. Color and color-pattern mechanisms of tiger beetles. *Ill. Biol. Monogr.* 3: 1–135.

Shelley, T. E., and D. L. Pearson. 1978a. Size and color discrimination of the robber fly *Efferia tricella* (Diptera: Asilidae) as a predator on tiger beetles (Coleoptera: Cicindelidae). *Environ. Entomol.* 7(6): 790–793.

———. 1978b. The attack response of *Efferia tricella* (Diptera: Asilidae) to eight tiger beetle species (Coleoptera: Cicindelidae). *J. N.Y. Entomol. Soc.* 85(4): 199.

Sherman, F., Jr. 1904. List of the Cicindelidae of North Carolina, with notes on the species. *Entomol. News* 15(1): 26–32.

Sherman, H. B. 1952. A list and bibliography of the mammals of Florida, living and extinct. *Quart. J. Florida Acad. Sci.* 15: 86–126.

Shields, O., and S. K. Dvorak. 1979. Butterfly distribution and continental drift between the Americas, the Caribbean, and Africa. *J. Nat. Hist.* 13: 221–250.

Shook, G. A. 1979. A note on a prey and predator of *Cicindela purpurea auduboni*. *Cicindela* 11(1): 12.

———. 1981. The status of the Columbia tiger beetle (*Cicindela columbica* Hatch) in Idaho (Coleoptera, Cicindelidae). *Pan-Pac. Entomol.* 57(2): 359–363.

———. 1984. Checklist of tiger beetles from Idaho (Coleoptera: Cicindelidae). *Great Basin Nat.* 44(1): 159–160.

———. 1989. *Cicindela lemniscata bajacalifornica*, a new subspecies of Cicindelidae (Coleoptera) from Baja California, Mexico. *Cicindela* 21(1): 1–11.

Shpeley, D. 1990. Ground beetle (Coleoptera: Cicindelidae and Carabidae) habitat selection at Andrew Lake, Alberta. *Provincial Museum of Alberta, Natural History Occasional Paper* 12: 47–62.

Simberloff, D. S. 1970. Taxonomic diversity of island biotas. *Evolution* 24 (1): 23–47.

———. 1976. Experimental zoogeography of islands: effects of island size. *Ecology* 57(4): 629–648.

Simpson, G. G. 1950. History of the fauna of Latin America. *Am. Sci.* 38: 361–389.

———. 1952a. Probabilities of dispersal in geological time. *Bull. Am. Mus. Nat. Hist.* 99: 113–176.

———. 1952b. The species concept. *Evolution* 5: 285–298.

Singh, T., and S. Gupta. 1982. Morphology and histology of the mandibular gland in *Cicindela sexpunctata* Fabr. (Coleoptera: Cicindelidae). *Uttar Pradesh J. Zool.* 2(1): 14–18.

Sloane, T. G. 1906. Revision of the Cicindelidae of Australia. *Proc. Linn. Soc. N.S.W.* 31: 309–360.

Slosson, A. T. 1895. Coleoptera at Lake Worth, Florida. *Can. Entomol.* 27: 9–10.

Smith, D. G. 1977. Spatial relationships of cicindelids. *Cicindela* 9(2): 29–33.

———. 1981. Screech owl predation on *Cicindela repanda* Dejean. *Cicindela* 13(3–4): 43–44.

———. 1988. Habitat selection in *Cicindela repanda*. *Cicindela* 20(2): 35–39.
Smith, D. G., and H. S. Gersham. 1979. Spatial distribution and grouping of cicindelids. *Cicindela* 11(4): 57–60.
Smith, H. M. 1969. Parapatry; sympatry or allopatry? *Syst. Zool.* 18(2): 254–255.
Smith, H. M., and F. N. White. 1956. A case for the trinomen. *Syst. Zool.* 5: 183–190.
Smith, P. W. 1957. An analysis of post-Wisconsin biogeography of the prairie peninsula region based on distributional phenomena among terrestrial vertebrate populations. *Ecology* 38: 205–213.
Smith, S. G., and R. S. Edgar. 1954. The sex-determining mechanism in some North American Cicindelidae (Coleoptera). *Rev. Suisse Zool.* 61: 657–667.
Smyth, E. G. 1935. Analysis of the *C. purpurea* group. *Entomol. News* 46: 14–19, 44–49.
Snow, F. H. 1878. *Amblycheila cylindriformis* Say. *Trans. Kans. Acad. Sci.* 6: 30–32.
Soans, A. B., and J. S. Soans. 1971. A convenient method of collecting the larvae of tiger beetles (order Coleoptera—family Cicindelidae) in the field. *Bombay Nat. Hist. Soc. J.* 68(2): 479–480.
———. 1972. A convenient method of rearing tiger beetles (Coleoptera: Cicindelidae) in the laboratory for biological and behavioural studies. *Bombay Nat. Hist. Soc. J.* 69(1): 209–210.
Spangler, H. G. 1988. Hearing in tiger beetles (Cicindelidae). *Physiol. Entomol.* 13(4): 447–452.
Spomer, S. M., and L. G. Higley. 1993. Population status and distribution of the salt creek tiger beetle, *Cicindela nevadica lincolniana* Casey (Coleoptera: Cicindelidae). *J. Kans. Entomol. Soc.* 66(4): 392–398.
St. John, E. P. 1936. Rare ferns of central Florida. *Am. Fern J.* 26: 41–55.
Stamatov, J. 1972. *Cicindela dorsalis* Say endangered on northern Atlantic coast. *Cicindela* 4(4): 78.
Stebbins, G. L., Jr. 1942. The role of isolation in the differentiation of plant species. *Biol. Symp.* 6: 217–233.
Stokes, W. L. 1973. *Essentials of earth history. An introduction to historical geology.* 3d ed. Englewood Cliffs, N.J.: Prentice Hall. 532 pp.
Sumlin, W. D. III. 1976a. A new subspecies of *Cicindela politula* from west Texas and a note on *Cicindela cazieri* (Coleoptera: Cicindelidae). *J. Kans. Entomol. Soc.* 49: 521–526.
———. 1976b. Notes on the tiger beetles of the genus *Cicindela* holdings of the Nevada State Department of Agriculture (Coleoptera: Cicindelidae). *Coleopt. Bull.* 30(1): 101–106.
———. 1982. Tiger hunting in the southwestern United States. *Environment Southwest* 25: 496.
———. 1985. A review of *Cicindela politula* LeConte (Coleoptera: Cicindelidae). *J. Kans. Entomol. Soc.* 58(2): 220–227.

Summers, S. V. 1873. List of Coleoptera of St. Louis County, Missouri. *Can. Entomol.* 5(7): 132–134.

Surkiewicz, J. 1993. Beetle-mania. *Nature Conservancy* 43: 32.

Swain, F. M. 1949. Upper Jurassic of northeastern Texas. *Am. Assoc. Petrol. Geol. Bull.* 33: 1206–1250.

Swan, L. A. 1975. Tiger beetle! [*Cicindela sexguttata*]. *Insect World Dig.* 2(1): 1–5.

Tanner, O. 1988. Of tiger beetles and wedge mussels: protecting Connecticut River riches. *Nature Conservancy News* 38(5): 4–11.

Tanner, V. M. 1927a. A preliminary study of the genitalia of female Coleoptera. *Trans. Entomol. Soc. Am.* 52: 5–50.

———. 1927b. A preliminary study of the genitalia of female Coleoptera. *Ann. Entomol. Soc. Am.* 15: 328–345.

———. 1929. The Coleoptera of Utah—Cicindelidae. *Pan-Pac. Entomol.* 6(2): 78–87.

Thompson, W. C. 1915. *Cicindela unipunctata* Fabr. at Seaville, New Jersey (Col.). *Entomol. News* 26: 425–426.

Thorpe, W. H. 1930. Biological races on insects and allied groups. *Biol. Rev. Camb. Philos. Soc.* 5: 177–212.

———. 1945. The evolutionary significance of habitat selection. *J. Anim. Ecol.* 14: 67–70.

Thurow, G. R. 1961. A salamander color variant associated with glacial boundaries. *Evolution* 15(3): 281–287.

Toh, Y., and A. Mizutani. 1994. Structure of the visual system of the larva of the tiger beetle (*Cicindela chinensis*). *Cell Tissue Res.* 278(1): 125–134.

Toulmin, L. D. 1955. Cenozoic geology of southeastern Alabama, Florida, and Georgia. *Am. Assoc. Pet. Geol. Bull.* 39: 207–235.

Tralau, H. 1967. The phytogeographic evolution of the genus *Ginko*. *Bot. Not.* 120: 409–422.

Tuomikoski, R. 1967. Notes on some principles of systematics. *Ann. Entomol. Fenn.* 33: 137–147.

Tuxen, S. L., ed. 1970. *Taxonomists glossary of genitalia in insects.* 2d ed. Darien, Conn.: S.H. Service Agency. 359 pp.

Uchupi, E. 1973. Eastern Yucatan continental margin and western Caribbean tectonics. *Am. Assoc. Pet. Geol. Bull.* 57: 1075–1085.

Uchupi, E., and K. O. Emery. 1968. Structure of continental margin off Gulf Coast of the United States. *Am. Assoc. Pet. Geol. Bull.* 52: 1162–1193.

Uchupi, E., and J. D. Milliman. 1971. Structure and origin of southeastern Bahamas. *Am. Assoc. Pet. Geol. Bull.* 55(5): 687–704.

Ulke, H. 1902. List of the beetles of District of Columbia. *Proc. U.S. Natl. Mus.* 25: 1–57.

Uscian, J. M., J. S. Miller, R. W. Howard, and D. W. Stanley Samuelson. 1992. Arachidonic and eicosapentaenoic acids in tissue lipids of two species of predacious insects, *Cicindela circumpicta* and *Asilis* sp. *Comp. Biochem. Physiol. B Comp. Biochem.* 103(4): 833–838.

Uscian, J. M., J. S. Miller, G. Sarath, and D. W. Stanley Samuelson. 1995. A digestive phospholipase A2 in the tiger beetle *Cicindela circumpicta*. *J. Insect. Physiol.* 41(2): 135–141.

U.S. Fish and Wildlife Service. 1990. Endangered and threatened wildlife and plants; determination of threatened status for the puritan tiger beetle and the northeastern beach tiger beetle. *Federal Register* 55: 32088–32094.

Valenti, M. A. 1986. Cicindelidae in the museum at the State University of New York College of Environmental Science and Forestry, Syracuse, New York. *Cicindela* 18(1): 1–6.

———. 1994. Two records of spiders preying on tiger beetles in the genus *Cicindela* (Coleoptera: Cicindelidae). *Cicindela* 26(2): 17–21.

Valenti, M. A., and F. E. Kurczewski. 1987. External morphology of adult *Cicindela repanda* Dejean. *Cicindela* 19(3/4): 37–49.

Valentine, B. D. 1947. Cicindelidae collected in Texas. *Coleopt. Bull.* 1: 61–62.

Valentine, J. M. 1945. Speciation and raciation in *Pseudanophthalmus* (Cavernicolous Carabidae). *Trans. Conn. Acad. Art. Sci.* 36: 631–672.

Van Dyke, E. C. 1929. The influence which geographical distribution has had in the production of the insect fauna of North America. *Trans. 4th Int. Congr. Entomol.* 555–566.

Van Natto, C., and R. Freitag. 1986. Solar radiation reflectivity of *Cicindela repanda* and *Agonum decentis* (Coleoptera: Carabidae). *Can. Entomol.* 118(2): 89–95.

Van Valen, L. 1965. Morphological variation and width of ecological niche. *Am. Nat.* 99: 377–390.

Varas, E. A. 1927. Third contribution to the study of the Cicindelidae group *purpurea—oregona*. *Rev. Chil. Hist. Nat.* 31: 173–175.

———. 1928. Fourth contribution to the study of Cicindelidae, group *formosa, purpurea, oregona*. *Rev. Chil. Hist. Nat.* 32: 231–250.

———. 1929. Contributions to the study of Cicindelidae, group *oregona*. *Rev. Chil. Hist. Nat.* 33: 394–402.

Vaurie, P. 1950a. Four new subspecies of the genus *Cicindela* (Coleoptera, Cicindelidae). *Am. Mus. Novit.* 1458: 1–6.

———. 1950b. Notes on the habitats of some North American tiger beetles. *J. N.Y. Entomol. Soc.* 58: 143–153.

———. 1951. Five new subspecies of tiger beetles of the genus *Cicindela* and two corrections (Coleoptera: Cicindelidae). *Am. Mus. Novit.* 1479 : 1–12.

———. 1955. A review of the North American genus *Amblycheila* (Coleoptera, Cicindelidae). *Am. Mus. Novit.* 1724: 1–26.

Vermeulen, H. J. W. 1993. The composition of the carabid fauna on poor sandy road-side verges in relation to comparable open areas. *Biodivers. Conserv.* 2(4): 331–350.

Vernon, R. O. 1951. Geology of Citrus and Levy Counties, Fla. *Fla. Geol. Surv. Geol. Bull.* 33: 1–256.

Vick, K. W., and S. J. Roman. 1985. Elevation of *Cicindela nigrior* to species rank. *Insecta Mundi* 1(1): 27–28.

Viedma, M. G., and M. L. Nelson. 1976. Significance of the morphological characters used in higher level natural classification of Coleoptera larvae. *Entomol. News* 87(9–10): 249–255.

Vogler, A. P. 1994. Extinction and the formation of phylogenetic lineages: diagnosing units of conservation management in the tiger beetle *Cicindela dorsalis*, pp. 261–273. *In:* B. Schierwater, B. Streit, G. P. Wagner, and R. DeSalle, eds., *Molecular ecology and evolution: approaches and applications.* Basel, Switzerland: Birkhauser Verlag.

Vogler, A. P., and R. Desalle. 1993. Phylogeographic patterns in coastal North American tiger beetles (*Cicindela dorsalis* Say) inferred from mitochondrial DNA sequences. *Evolution* 47(4): 1192–1202.

———. 1994a. Mitochondrial DNA evolution and the application of the phylogenetic species concept in the *Cicindela dorsalis* complex (Coleoptera: Cicindelidae). *Ser. Entomol.* 51: 79–85.

———. 1994b. Diagnosing units of conservation management. *Conserv. Biol.* 8(2): 354–363.

———. 1994c. Evolution and phylogenetic information content of the ITS-1 region in the tiger beetle *Cicindela dorsalis. Mol. Biol. Evol.* 11(3): 392–405.

———. 1994d. Phylogeographic patterns in coastal North American tiger beetles (*Cicindela dorsalis* Say) inferred from mitochondrial DNA sequences. *Evolution* 47(4): 1192–1202.

Vogler, A. P., R. Desalle, T. Assmann, C. B. Knisley, and T. D. Schultz. 1993. Molecular population genetics of the endangered tiger beetle *Cicindela dorsalis* (Coleoptera: Cicindelidae). *Ann. Entomol. Soc. Am.* 86(2): 142–152.

Vogler, A. P., C. B. Knisley, S. B. Glueck, J. M. Hill, and R. Desalle. 1993b. Using molecular and ecological data to diagnose endangered populations of the puritan tiger beetle *Cicindela puritana. Mol. Ecol.* 2(6): 375–383.

Vogt, G. B. 1949. Three new Cicindelidae from south Texas with collecting notes on other Cicindelidae (Coleoptera). *Bull. Brooklyn Entomol. Soc.* 44(1): 1–9.

Wagenaar Hummelinck, P. 1955. Caribbean tiger beetles of the genus *Megacephala. Studies on the fauna of Curaçao and other Caribbean Islands* 6(28): 89–125.

———. 1983. Additional notes on Caribbean tiger-beetles of the genera *Cicindela* and *Megacephala. Uitgaven Natuurwetenschappelijke Studiekring voor Suriname en de Nederlandse Antillen* 111: 69–135.

Wallace, F. L., and R. C. Fox. 1975. A comparative morphological study of the hind wing venation of the order Coleoptera, Part 1. *Proc. Entomol. Soc. Wash.* 77 (3): 329–354.

Wallis, J. B. 1961. *The Cicindelidae of Canada*. Toronto: University of Toronto Press. 74 pp.

Ward, R. D. 1971. Spring collecting in Louisiana, Texas and Arkansas with some notes on *Cicindela sexguttata tridens* Casey. *Cicindela* 3(4): 69–77.

———. 1980. Lectotype designations for the species of tiger beetles described by George Henry Horn. *Omus* spp., *Cicindela* spp. *Cicindela* 12(4): 57–63.

———. 1989. The history and status of three cicindelid names attributed to George Henry Horn (Coleoptera: Cicindelidae). *Cicindela* 21(3/4): 33–39.

Ward, R. D., and T. A. Bowling. 1980. *Cicindela* collected from malaise traps in Michigan and notes on the distribution of Michigan species. *Cicindela* 12(2): 29–31.

Watrous, L. E., and Q. D. Wheeler. 1981. The out-group comparison method of character analysis. *Syst. Zool.* 30(1): 1–11.

Watts, W. A. 1969. A pollen diagram from Mud Lake, Marion County, north central Florida. *Geol. Soc. Am. Bull.* 80: 631–642.

———. 1970. The full glacial vegetation of north-western Georgia. *Ecology* 51: 17–33.

———. 1973. The vegetation record of a mid-Wisconsin interstadial in northwest Georgia. *Quat. Res.* 3: 257–268.

Webb, W. L. 1950. Biogeographic regions of Texas and Oklahoma. *Ecology* 31: 426–433.

Weidie, A. E., and G. E. Murray. 1967. Geology of Parras Basin, and adjacent center of northeastern Mexico. *Am. Assoc. Pet. Geol. Bull.* 51(5): 678–695.

Werner, F. 1925. Zur kenntnis der Fauna der insel Bonaire. *Z. Wiss. Zool.* 125: 533–695.

Wheeler, Q. D. 1979. Revision and cladistics of the Middle American genus *Creagrophorus* Matthews (Coleoptera: Leiodidae). *Quaest. Entomol.* 15: 447–479.

Whitehead, D. R. 1972a. Classification, phylogeny, and zoogeography of *Schizogenius* Putzeys (Coleoptera: Carabidae: Scaritini). *Quaest. Entomol.* 8: 131–348.

———. 1972b. Development and environmental history of the Dismal Swamp. *Ecol. Monogr.* 42: 301–315.

Whitehead, D. R., and G. E. Ball. 1975. Classification of the middle American genus *Cyrtolaus* Bates (Coleoptera: Carabidae: Pterostichini). *Quaest. Entomol.* 11: 591–619.

Whittaker, R. H. 1956. Vegetation of the Great Smoky Mountains. *Ecol. Monogr.* 26: 1–80.

Wickham, H. F. 1894. The Coleoptera of Canada. The Cicindelidae of Ontario and Quebec. *Can. Entomol.* 26(6): 149–154.

———. 1899. Habits of North American Cicindelidae. *Proc. Davenport Acad. Nat. Sci.* 7: 206–228.

———. 1902. A catalogue of the Coleoptera of Colorado. *Bulletin from the Laboratories of Natural History of the State University of Iowa* 5: 217–310.

———. 1904. The influence of the mutations of the Pleistocene lakes upon the present distribution of *Cicindela*. *Am. Nat.* 38: 643–654.

Wiesner, J. 1981. Eine neue *Oxychila* aus Mexiko (Col., Cicindelidae). *Entomol. Basiliensia* 6: 150–153.

Wilhelm, O., and M. Ewing. 1972. Geology and history of the Gulf of Mexico. *Bull. Geol. Soc. Am.* 83: 575–600.

Will, K. W., F. F. Purrington, and D. J. Horn. 1995. Ground beetles of islands in the western basin of Lake Erie and the adjacent mainland (Coleoptera: Carabidae, including Cicindelini). *Gt. Lakes Entomol.* 28(1): 55–70.

Williams, F. X. 1916. Notes on the life history of *Methoca stygica* Say. *Psyche* 23: 121–125.

———. 1928. *Pterombrus,* a wasp-enemy of the larva of tiger beetles. *In:* F. X. Williams, Studies in tropical wasps: their hosts and associates (with descriptions of new species). *Hawaii Sugar Plant. Assoc. Entomol. Bull.* 19: 144–151.

Williams, I. W. 1938. The comparative morphology of the mouthparts of the order Coleoptera treated from the standpoint of phylogeny. *J. N.Y. Entomol. Soc.* 46: 145–267.

Williamson, J. D. M. 1959. Gulf Coast Cenozoic history. *Trans. Gulf Coast Assoc. Geol. Soc.* 9: 15–29.

Willis, H. L. 1967. Bionomics and zoogeography of tiger beetles of saline habitats in the central United States (Coleoptera: Cicindelidae). *Univ. Kans. Sci. Bull.* 47: 145–313.

———. 1968. Artificial key to the species of *Cicindela* of North America north of Mexico (Coleoptera: Cicindelidae). *J. Kans. Entomol. Soc.* 41: 303–317.

———. 1969. Translation and condensation of Horn's key to world genera of Cicindelidae. *Cicindela* 1(3): 1–16.

———. 1970. The Cicindelidae of Kansas. *Cicindela* 2: 1–27.

———. 1972. Species density of North American *Cicindela*. *Cicindela* 4: 29–34.

———. 1980. Description of the larva of *Cicindela patruela*. *Cicindela* 12(4): 49–56.

Willis, H. L., and J. Stamatov. 1971. Collecting Cicindelidae in the Northwest. *Cicindela* 3(3): 41–51.

Wilson, D. A. 1970. Three subspecies of cicindelids threatened with extermination. *Cicindela* 2: 18–20.

———. 1974. Survival of cicindelid larvae after flooding. *Cicindela* 6(4): 79–82.

———. 1975. Cicindelid collecting in the southern United States. *Cicindela* 7(3): 41–51.

———. 1978. The tiger beetles of Mount Desert Island and Acadia National Park, Maine. *Cicindela* 10(1): 7–16.

———. 1979. *Cicindela ancocisconensis* T.W. Harris. *Cicindela* 11(3): 33–48.

Wilson, D. A., and A. E. Brower. 1983. The Cicindelidae of Maine. *Cicindela* 15(1–4): 1–33.

Wilson, D. S. 1978. Prudent predation: a field study involving three species of tiger beetles. *Oikos* 31(1): 128–136.

Wilson, E. O., and W. L. Brown. 1953. The subspecies concept and its taxonomic application. *Syst. Zool.* 2(3): 97–111.

Wilson, E. O., and D. J. Farish. 1973. Predatory behavior in the ant-like wasp *Methoca stygica* (Say) (Hymenoptera: Tiphiidae). *Anim. Behav.* 21: 292–295.

Wise, K. A. J. 1990. Notes on coastal tiger beetles (Coleoptera: Cicindelidae). *Records of the Auckland Institute and Museum* 27: 181–184.

Wolcott, G. N. 1936. "Insectae Borinquenses." A revision of "Insectae Portoricensis." A preliminary annotated check-list of the insects of Porto Rico, with descriptions of some new species. *J. Agric. Univ. P.R.* 20: 1–601.

———. 1948. The insects of Puerto Rico. Coleoptera. *J. Agric. Univ. P.R.* 32(2): 225–416.

Woodring, W. P. 1929. Tectonic features of the Caribbean region. *Proceedings, 3d Pan-Pacific Science Congress* 1: 401–431.

———. 1954. Caribbean land and sea through the ages. *Bull. Geol. Soc. Am.* 65: 719–732.

Woodruff, R. E., and R. C. Graves. 1963. *Cicindela olivacea* Chaudoir, an endemic Cuban tiger beetle, established in the Florida Keys (Coleoptera: Cicindelidae). *Coleopt. Bull.* 17: 79–83.

Wray, D. L. 1950. Insects of North Carolina. Second Supplement. Raleigh: North Carolina Department of Agriculture, Division of Entomology. 59 pp.

Wright, H. E., Jr. 1972. Interglacial and post glacial climates, the pollen record. *Quat. Res.* 2: 274–282.

Yager, D. D., and H. G. Spangler. 1995. Characterization of auditory afferents in the tiger beetle, *Cicindela marutha* Dow. *J. Comp. Physiol. A Sens. Neural Behav. Physiol.* 176(5): 587–599.

Yarbrough, W. W., and C. B. Knisley. 1994. Distribution and abundance of the coastal tiger beetle, *Cicindela dorsalis media* (Coleoptera: Cicindelidae), in South Carolina. *Entomol. News* 105(4): 189–194.

Young, O. P. 1980. Predation by tiger beetles (Coleoptera: Cicindelidae) on dung beetles (Coleoptera, Scarabaeidae) in Panama. *Coleopt. Bull.* 34(1): 63–65.

Zikan, J. J. 1929. Zur Biologie der Cicindeliden Braziliens. *Zool. Anz.* 82: 269–414.

Index

Scientific names are italicized and listed in alphabetical order. Bold-faced type indicates pages containing plates, maps, or figures for a particular species or subject.

Amblycheila, 43
Aneides, 26–27
Aphelocoma coerulescens, 31, 32
Aphelocoma ultramarina, 32
Apical lunule, **51**
Asilidae, 4. *See also* robber flies

Basal dot, **50**
Bee flies, 4
Black widow spider, *Latrodectus,* 30–31
Bombyliidae. *See* Bee flies

Calcaritermes, 30
Cicindela. See individual species
C. abdominalis, 13, 31, 55, **82–83, 126**
C. ancocisconensis, 13, 59, **80, 125**
C. blanda, 14, 50, 56, **107–108, 130**
C. cuprascens, 14, 56, **108, 131**
C. cursitans, 14, 57, **104, 130**
C. dorsalis media, xviii, 14, 22, 23, 54, **94–95, 128**
C. dorsalis saulcyi, xviii, 14, 54, **96–97, 128**
C. duodecimguttata, 13, 14, 58, **65, 121**
C. formosa generosa, 14, 49, 59, **66–67, 121**
C. gratiosa, 1, 13, 55, **109–110, 132**
C. hamata lacerata, 14, 15, 49, 56, **110–111, 133**
C. highlandensis, xviii–xix, 13, 36, 55, **84–85, 126**
C. hirticollis, 14, 23, 51, 60, **67–68, 122**
C. hirticollis rhodensis, **68**
C. hirtilabris, 13, 36, 55, **112–113, 132**
C. lepida, 14, 56, **114, 132**
C. limbalis, 14, 59, **69, 123**
C. longilabris, 14, 57, **70, 123**
C. macra, 14, 56, **114, 131**
C. marginata, 14, 15, 22, 23, 50, 56, **115–116, 132–133**
C. marginipennis, 13, 55, **85–86, 126**
C. nigrior, 13, 59, **71, 124**
C. olivacea, xix, 14, 35, 58, **101–102, 129**
C. patruela, 14, 57, **72, 123**
C. pilatei, 52, 57, **106, 130**
C. punctulata, 13, 58, **87, 127**
C. puritana, 13, 56, **116–117, 131**
C. purpurea, 14, 59, 60, **72, 122**
C. repanda, 13, 51, 58, **73, 124**
C. rufiventris rufiventris, 14, 55, **89–90, 127**
C. rufiventris hentzii, 15, **88–89, 127**
C. scabrosa, 13, 31, 36, 50, 51, 55, **91–92, 126**
C. scutellaris lecontei, 14, **75**
C. scutellaris unicolor, 13, 49, 51, 57, 59, **74–75, 125**
C. severa, 15, 58, **98, 129**
C. sexguttata, 14, 15, 51, 58, **76–77, 123–124**
C. splendida, 14, 59, **78–79, 122**
C. striga, 15, 57, **99–100, 129**
C. togata, 15, 55, **100–101, 128**
C. tranquebarica, 14, 59, 60, **81, 125**

C. trifasciata ascendens, 14, 15, 58, **93**, 127
C. unipunctata, 14, 57, **105**, 130
C. viridicollis, xx, 14, 35, 57, **103–104**, 129
C. wapleri, 14, 56, **118**, 131
Cicindelidae. *See* Tiger beetles
Cicindelini, 44
Clypeus, 2, 47
Coastal dunes species, 14
Cobblestone tiger beetle, 10. *See C. marginipennis*
Collecting and preserving tiger beetles: adults, 4–5; larvae, 6
Collyrini, 3
Coral-inhabiting species, 14
Croizat, 26
Ctenostomatini, 3
Cuban tiger beetles, 35. *See also C. olivacea; C. viridicollis*

Decumbent setae, **52**
Diptera, 4
Dragonfly, 4

Ecology of tiger beetles, 2–3
Endangered species. *See C. puritana; C. highlandensis*
Equipment for photographing tiger beetles, **6–8**

Femur, **50**
Ferns of central Florida, 27
Florida counties, map of, **61**
Florida, geological history: relationship to distribution patterns, 24–36
Florida highlands tiger beetle, 11. *See C. highlandensis*
Frons, 47

Gena, **49**
Geological timescale, 36
Granular elytra, **51**
Gravel pits, 14
Glyptotermes, 30

Habitat destruction, xviii–xix; by vehicular traffic, xix
Habitat segregation of species, 22–23
Habitats of eastern tiger beetles, **12–22**
Humeral lunule, 51
Hyla femoralis, 34

Identification keys: Nearctic genera, 42; eastern genera, 53; species of *Megacephala*, 53–54; species of *Cicindela*, 54–60

Labels, collection of, 5
Labrum, **47**; medial teeth, **48**
Larvae of tiger beetles, 3, 19
Larvae, *C. dorsalis media*, 94–95
Latrodectus, 30–31
Lepidoptera, 26
Locating and photographing tiger beetles, 6
Lunule. *See* apical lunule; humeral lunule

Mandible, *C. marginata*, **48**
Mandibular tooth, **48**
Mantica, 3
Manticora, 3
Marginal line, **51**
Megacephala, 43, 52, **119**
M. carolina carolina, 14, 54, **62–63**, **119**
M. carolina chevrolati, 45, 54, 62
M. carolina floridana, 15, 54, **63**, **119**
M. virginica, 15, 53, **64**, **119**
Megacephalini, 43
Mesosternum, **51**
Metasternum, **51**
Microhyla carolinensis, 34
Microserrulation, **50**
Middle band, **51**
Morphological characters, 47, **48–52**
Morphology, characters used for identification, 48–52
Mycotrupes, 24–25

Omus, 43
Ophioglossum, 27

Ophisaurus, legless lizards, 35
Orthoptera, 29–30

Paratettix, 29–30
Polygenis, 28
Peromyscus, 28–29, 33, 35
Photography techniques, equipment, **7–8**
Predators and parasites of tiger beetles, 3
Pronotum, **50**
Prorhinotermes, 30
Prosternum, **51**
Puritan tiger beetle, 10. See also *C. puritana*

Robber flies, 4

Salamanders, 26–27
Scrub Jay, *Aphelocoma coerulescens*, 31–33
Sossipus, 31
Species criteria, 37–40
Species list, 45–46
Species restricted to Florida, 36. See also *C. highlandensis*; *C. hirtilabris*; *C. scabrosa*
Species segregation, 22–23
Subspecies, 39
Sutural spine, **49, 50**
Suwannee Strait, 25

Tarsus, **50**
Termites, 30
Tetracha, 43
Tibia, **50**
Tiger beetles: defined, 1–2; ecology of, 2–3; larvae of, 3; predators and parasites of, 3–4; collecting and preserving of, 4–5; degreasing specimens of, 5; collection label, 5; collecting and preserving larvae of, 6; locating, 6; photographing, **6–8**; studying, 9–10; conservation issues, 10–11; restriction to Florida, 36; classification of, 41–44; list of eastern species of, 45–46
—habitats of: roadside, 13, **15, 16**; scrub, 13, **16**; rivers and streams, 13, **17, 18**; seashores, 14, **18, 19**; coral, 14, **19**; gravel pits, 14, **20**; woodland paths, 14, **21**; eroded banks, 14, **21**; salt marshes, 15, **21**; coastal alkali mudflats, 15, **22**; rock outcroppings, 15, **22**
Tiphiidae, 4
Treefrog, *Hyla*, 34–35

Vehicle, beach. See habitat destruction

What is a tiger beetle, 1–2
Why study tiger beetles, 9

Paul M. Choate is a lecturer in the Department of Entomology and Nematology at the University of Florida, Gainesville, where he teaches insect classification and an online computer course. He has published taxonomic papers on tiger beetles and ground beetles, including the description of a new species of tiger beetle from Florida. He also specializes in field photography of tiger beetles.